Learning from the Past, Looking into the Future: Realism in Philippine Foreign Policy

Generoso D.G. Calonge

Ambassador of the Republic of the Philippines
(Retired)

Published by Broken Column Press, LLC
BrokenColumnPress.com
ISBN: 978-1-944616-20-5
LCCN: 2024900609

TABLE OF CONTENTS

Dedication

This humble work is dedicated to the Filipino people. May their government take care of their territory and resources so that the future generations can look back and feel proud of their national heritage. Filipinos are entitled to security, the highest, and most sublime, responsibility of the State.

Learning from the Past, Looking into the Future

Preface

This book tries to highlight what appears to the author to be some of the shortcomings of the foreign policy approach of the Philippines. The author is of the view that two considerations in decision-making are deficient: national interest and state power. Through this work, the author makes suggestions, based on his background and experience in government.

On the national interest, the country started out with a huge swathe of territory bequeathed by the Treaty of Paris between the United States and Spain in 1898. The country's leaders did not cement ownership of that wide expanse of water. On state power, the Philippines did not prepare itself so that it could secure its territory after independence in 1946. While it is true that it was beset with internal dissidence, the external defense mission was way short of fulfillment under the mandatory portfolio of nationhood. For this neglect, the Philippines paid the price in that it lost territory, including access to resources on and beneath water, that it was entitled to under international law.

Given the above-mentioned historical background, the author suggests that Philippine leaders, diplomats and shapers of public opinion need to shift their focus of analysis from the current naive approach of legalism to realism. Realism is at the base of international political analysis in the sense that its approach precisely takes care of what the Philippines needs as a theoretical framework. It was codified by Morgenthau, systematized by Waltz, and applied by practitioners like Kennan and Kissinger.

The loss of territory and territorial rights is one of the most devastating blows that a nation can suffer. Never again should the Philippines bear such a catastrophe. It is time to mature as a nation.

Leadership is key to the renaissance of the Philippine psyche. The future generations of Filipinos will depend on two vital elements that their leaders should concentrate on: the appropriate foreign and defense policies consistent with the national interest and the required state power. These elements have to be underpinned by a robust economy to support expenditures and investments in them.

The Philippines has to be equal to the task if it seeks to deserve to be a respectable state in the community of nations.

Generoso D.G. Calonge

The first duty of the state is to defend itself, and outside of a juridical order none but the state itself can define the actions required.

Kenneth N. Waltz
Structural Realism After the Cold War
International Security, Summer, 2000, Vol. 25, No. 1

Generoso D.G. Calonge

Chapter 1: The Problem and Background of the Study

Introduction

The perceived gap. This work aspires to investigate the role of realism in Philippine foreign policy. Realism is a paradigm in international relations that gives emphasis on the power of the state within a system that has no central authority. In other words, each state works on its own, considering its national interest in utmost competition with the others. The author is of the view that Philippine foreign policy could be analyzed from the prism of classical realism (or one of its variants, as applicable, like neorealism) to determine whether the Philippine state puts its national interest and its quest for power on top of its priorities.

The desire to closely investigate the Philippines' position internationally is meant to improve its approaches in comparison with other nations that appear more successful in the pursuit of their interests. A case can be made that the stature of the Philippines a few decades before and after the Second World War was in a much enviable position. The impression is that the Philippines deteriorated in its approach to national challenges, including that of foreign policy.

Background of the Study

The author wants to determine whether the Philippine state puts its national interest and its quest for power on top of its priorities. The political milieu within which the Philippines operates may provide the answers to this question. For example, the executive branch of the government, as well as the legislature, in terms of laws passed, and the judiciary could be fertile grounds for the responses to the issue. Policy pronouncements of the government and speeches of leaders as relayed by the media are important components of the analysis. It is the intention of the author to cover news releases as a reflection of prevailing opinion or sentiment on the way national interest and power are projected.

This study sprang out of almost forty-four years of government service of the author. Questions have emerged from within his own reflections why there is a need to scientifically define where lies the Philippines in foreign policy making. This query would be followed by the ultimate test: is its foreign policy consistent with its needs, its ambitions, and its rightful place in the family of nations.

There are many schools of thought upon which analysis can be based in studying the foreign relations positions of countries. Some of these are: liberalism, rational choice, constructivism, classical realism and neorealism. The author argues that classical realism remains a valid yardstick to assess the performance of the Philippines in international affairs. In fact, it can be argued that realism is the most appropriate prism from which conclusions about Philippine foreign policy making can be drawn.

The author bases his contentions around the principles espoused by Hans Morgenthau in his classic book, *Politics Among Nations*[1]. In his work, national interest and power, as well as self-preservation, are the cornerstones of a country's foreign policy.

A Look at Some Theories in International Relations

The author argues that some of the schools of thought, or philosophical theories in international relations, namely: Idealism, Liberalism, Rational Choice, Constructivism and Realism have their own strengths. Realism, however, is arguably the most basic of all analytical tools. Let us take them up in turn.

---

[1]Morgenthau, 1948.

Idealism. A rectorial address defined idealism as "the spirit which impels an individual or group of individuals to a loftier standard of conduct than that which ordinarily prevails around him or them."[2] In the international field, the Earl of Birkenhead added that "Idealism is the spirit which would carry into the relations of States the kind of ethical progress generally indicated above."[3] As we can see, idealism abhors unpleasant behavior. In this context, we can include wars, poverty, injustice and other manifestations of hardship that idealists seek to eliminate.

Contrasted with realism, idealism, according to one scholar, may come in two senses, "it may be highly praised as the expression of the best elements in the *zoon politicon* (political animal), or it may be ridiculed as the naive belief of a political fool."[4]

Liberalism. Liberalism seeks "to construct institutions that protect individual freedom by limiting and checking political power. While these are issues of domestic politics, the realm of International Relations (IR) is also important to liberals because a state's activities abroad can have a strong influence on liberty at home."[5]

Liberalism is identified with issues such as feminism, diversity, gender sensitivity, inclusiveness and others that used to be known around the fringes of so-called mainstream studies. A

---

[2] *Lecture by the Earl of Birkenhead,* **Lord Rector of Glasgow University,** November 3, 1923, p. 1.

[3] Ibid.

[4] Mordecai Roshwald, *Realism and Idealism in Politics,* **Social Science,** Published by Pi Gamma Mu, International Honor Society in Social Sciences, April 1971, Vol. 46, No. 2, p. 100. https://www.jstor.org/stable/41885877.

[5] Jeffrey W. Meiser, *Introducing LIberalism in International Relations Theory,* February 18, 2018. https://www.e-ir.info/2018/02/18/ introducing-liberalism-in-international-relations-theory.

case has quite convincingly been put forward that integration of minorities leads to progress.[6]

Liberalism seeks equality with other theories in the sense that it should be accorded the same level of respect. Moravcsik says that liberalism has three core assumptions, enumerated as follows:

1. The primacy of societal actors, including individuals and private groups who are normally rational and risk averse;

2. Representation and state preferences whereby the state represents a subset of domestic society, on the basis of whose interests state officials define state preferences and act purposively in world politics;

3. Interdependence and the international system: the configuration of interdependent state preferences determines state behavior.[7]

Rational Choice. This theory presupposes that an individual or a leader will make calculations based on the cost and benefit analysis of any situation or challenge and decide accordingly. Rationalism makes it appear that the logical progression from the facts to the results are systematic. In fact, the process is undertaken within the context of "incomplete

---

[6] Dai Le and Katie Calvey, *Cultural Diversity in Politics and Media Will Create National Prosperity*, in Disruptive Asia, https://disruptiveasia.asiasociety.org/cultural-diversity-in-politics-and-media-will-create-national-prosperity. This study asserts that Australia is still short of the ideal figure for minority integration but it is keenly working on it.

[7] Andrew Moravcsik, *Taking Preferences Seriously: A Liberal Theory of International Politics*, **International Organization**, Autumn, 1997, Vol. 51, No. 4, pp. 516-521.

information" and "lack of credibility." Rational choice merely offers a method to explain social phenomena.[8]

Among and within the three branches of government, the rational choice theory also works, bureaucratically. The United States Constitution, from which our executive, legislative, and judicial branches have been patterned, also employs the rational choice theory, especially in foreign affairs and the management of national security. The executive branch has general primacy in international affairs. "The rational choice model confirms that the strategic dynamics among the three branches do operate to prevent any single entity from enjoying a monopoly of all government power."[9]

Constructivism. Constructivism sees the world according to the interpretation of people that is based on their relationships and interactions. One example given is the notion of the danger posed by nuclear weapons. The argument of constructivists is that hundreds of nuclear missiles would not be threatening to the Western powers but five nuclear bombs held by North Korea would create nervousness in the leaders of same countries. The atmospherics would be different. It is how one person or one country perceives reality. The social construct says everything.[10]

---

[8] Fearon, James; Wendt, Alexander, *Rationalism v. Constructivism: A Skeptical View,* **Handbook of International Relations**, SAGE (2002).

[9] John O. McGinnis, *Constitutional Review by the Executive in Foreign Affairs and War Powers: A Consequence of Rational Choice in the Separation of Powers,* **Law and *Contemporary Problems***, Autumn, 1993, Vol. 56, No. 4, Elected Branch Influences in Constitutional Decisionmaking (Autumn, 1993), pp. 293-325. Mr. McGinnis cited on page 324, "THE FEDERALIST No. 51, at 321-22 (James Madison) (Clinton Rossiter ed., 1961) (premise of separation of powers is that "ambition must be made to counteract ambition")."

[10] Sarina Theys, *Introducing Constructivism in International Relations Theory,* February 23, 2018.https://www.e-ir.info/2018/02/23/introducing-constructivism-in-international-relations-theory/.

Constructivism begins with mutually agreed common understanding upon which the basis of impressions, or even judgments, are made. These include identities under certain historical, cultural, political, and social contexts.[11] As a process, constructivism seeks patterns of behavior. It is quite ambitious as a theory in the sense that it wants to reach into the future on the basis of mentally or psychologically established interpretations, relationships, and perceptions. As Ted Hopf said: "In effect, the promise of constructivism is to restore a kind of partial order and predictability to world politics that derives not from imposed homogeneity, but from an appreciation of difference."[12]

Neorealism. Neorealism is a concept based on the assumption that nations belong to an international system. It is a scientific approach to international relations where, aside from anarchy, the distribution of power is important in explaining the role of states in their dealings with each other. Neorealism is also known as structural realism. It was founded by Kenneth Waltz, an American political scientist, who also added level of analysis as part of its repertoire in that nations could be looked upon as individual units.[13] In using the level of analysis, countries could be discrete elements at the unit level, while all countries can be viewed together as comprising the system level.

Neorealism aims to explain international relations as briefly, as comprehensively, and as substantively as possible. It has

---

[11] Ted Hopf, *The Promise of Constructivism in International Relations Theory*, **International Security**, The MIT Press, Summer, 1998, Vol. 23, No.1 (Summer, 1998), p. 176 https://www.jstor.org/stable/2539267.
[12] *Ibid.* p. 200.
[13] *Theory of International Politics*, 1979.

been described as "the most parsimonious, or least complex, theory of International Relations."[14]

Waltz, the founder of the theory, sought to establish an overarching discipline that can adequately explain politics among nations. He claimed that "Explanations of international politics are not to be found at the state or individual level of political decision makers, but at the level of the international system."[15] He contrasts domestic political structure and the international system. In domestic politics, hierarchy is the ordering principle; in international politics, anarchy is the ordering principle.[16] Thus, "the distribution of capabilities across units becomes a defining criterion for the structure of the system."[17] The distribution of power results on whether the system is *multipolar*, *bipolar*, or *unipolar*.[18]

Realism. Realism is the overarching framework of this book. The author chose classical realism because it was, in his view, the first comprehensive work on international relations that gained popularity worldwide. In *Politics Among Nations,* Morgenthau expounded on his six principles of realism, as follows:

- Human nature is the root of laws that govern politics

- National power is the yardstick of national interest

- National interest is always changing

- Politics has no regard for moral principles

---

[14] Manuela Spindler, *Neorealist Theory,* **International Relations A Self-Study Guide to Theory,** Published by: Verlag Barbara Budrich, 2013, p. 124 http://www.jstor.com/stable/j.ctvdf09vd.8.

[15] *Ibid.* p. 126.

[16] *Ibid.* p. 132.

[17] *Ibid.*

[18] *Ibid.*

- There is a difference between the moral aspirations of a nation and universal principles of morality

- International politics is autonomous[19]

Morgenthau firmly believed that the basic problems in international relations "stem from the very essence of human nature."[20] Such problems do not spring from some sudden developments or temporary configurations of the international order. One such characteristic of human nature is its never ending search for and achievement of power.

Realism relies on national interest in its analysis. It also uses the concept of "balance of power" among individual states to determine their status in a potential conflict. He justifies the use of balance of power as an analytical construct. He says that "The balance of power, you may say, is for foreign policy what the law of gravity is for nature; that is, it is the very essence of foreign policy."[21]

Indeed, the realist theory that Morgenthau has espoused is the grounding upon which basic interpretation of world events are founded. The much-sought desire of international relations theory to predict future events has been achieved by him, at least once. In referring to what is now the People's Republic of China as "Communists China," he emphatically said: "As long as Communist China remains a backward, underdeveloped nation, it is only a potential threat to the rest of the world. But once 600 million Chinese are in possession of the modern instruments of

---

[19] Hans J. Morgenthau, *Politics Among Nations: The Struggle for Power and Peace,* 5th ed.; Revised; (New York: Alfred A. Knopf, 1978), pp. 4-15.
[20] Hans J. Morgenthau, *Realism in International Politics,* **Naval War College Review**, Vol. 10, No. 5 (January, 1958), p. 1. Published by the U. S. Naval War College Press https://www.jstor.org/stable/44640810.
[21] *Ibid.* p. 5.

industry, then they will become an enormous threat to the rest of mankind, the Soviet Union included."[22] Those words were spoken in the late 1950s. Contemporary events have proven that Morgenthau was correct because later, in the 1960s, China and the Soviet Union had ideologically separated. China also opened up to the West, especially to the United States. But recent developments are showing that China and the United States do not enjoy harmonious relations on many issues: the South China Sea, Taiwan, trade, technology and even in the Olympics.[23]

The question: Is Philippine foreign policy realist enough?

The purpose of this book is to help unravel the dynamics of Philippine foreign policy formulation, whether it takes into account the best interests of the nation and its people, or whether there is a dearth of analyses in government actions to test the inclusion of the national interest into policy. The study strives to understand why Filipino leaders act the way they do vis-à-vis foreign issues. For example, why did the Philippines, as a state, do away with the Treaty of Paris in its boundaries? This treaty predated the UNCLOS,[24] and even all the prior negotiations that led to its conclusion. Was that action beneficial to Philippine interests?

More fundamentally, the Philippines has not lent coherence to very basic issues that concern the nation. One example is the definition of its territory. While the case of *Magallona v. The Executive Secretary* appears to have settled the matter by the country's compliance with the provisions of UNCLOS, there are lingering constitutional questions. The Treaty of Paris limits seems to the author to be more advantageous to the

---

[22] *Ibid.*

[23] The United States and many Western countries did not send official representatives to the recent Winter Olympics held in Beijing.

[24] *United Nations Convention on the Law of the Sea, 1982.*

Philippines. A former senior diplomat wrote that "The Philippines should make up its mind on the various controversial issues surrounding the definition of the Philippines' territory and maritime regimes and take definitive national positions on them."[25] Certainly, a poorly articulated, or undefined national territory, is not good for the national interest.

Significance of this Work

This book will try to apply the heavily theoretical nature of realist literature to a concrete case, and that is the Philippine experience. The author will provide a real-life example of the realist theory as applied on the ground. It can become a template in the analysis of countries that are similarly situated. It will refine the theory in the sense that it can show the conditions under which the classical realist proposition will or will not apply.

The question is worth exploring because of the role and significance of the Philippines in international affairs. The findings will matter as they provide guideposts for leaders and future generations of foreign policy watchers like government officials, academics, and diplomats, while engaged in shaping guidelines for the nation.

As a result of this study, theory and practice will be broadened, and even improved, in that there will arise an example of the application of the realist theory in real life, thereby enhancing the literature on the subject.

This Work's Scope and Limitations

An aspiration of realists is to attain a scientific status of international relations theory so that it will be able to predict

---

[25] Rodolfo C. Severino, *The Philippines' National Territory*, **Southeast Asian Affairs**, 2012, (2012) p. 257 Published by ISEAS–Yusof Ishak Institute, https://www.jstor.org/stable/41713998.

outcomes in approaches to foreign policy. This is too lofty a goal because it is unattainable. The scope of the study will be within what Morgenthau was said to have expressed "that we must see the world as it is, rather than as we want it to be."[26] The nearly hit-and-miss nature of describing trends in the future is an accepted fact in foreign policy analysis. One does not have to win them all. Getting some predictions to come true may be good enough, especially when the positive results turn out to be the big issues of the year.[27] Still, there are those who claim to have been prescient about the occurrence of the burning topics of the day. One author predicted that Russia would be provoked, leading to interpretation of unfriendly acts perceived by Russian leaders, once an eastward expansion of NATO was envisaged by the West. He said that he mentioned these words in 1994,[28] long before Russia's war against Ukraine. Another realist theoretician wrote that his article about the situation in Ukraine "should not be controversial, given that many American foreign policy experts have warned against NATO expansion since the late 1990s."[29] There is potential for predictability of world events after all, using the tools of analysis of realism.

Limitations of the study include the contextual nature of decision-making from the point of view of time and geopolitical changes. There is also the problem of individuals, especially the leaders, having to decide "on the basis of incomplete information, deal with unpredictability of their actions and cope with

---

[26] Samuel Barkin, *Realism, Prediction, and Foreign Policy*, **Foreign Policy Analysis** (2009) **5**, p. 233 https://www.jstor.org/stable/2490977.

[27] https://ecfr.eu/article/the-long-year-top-foreign-policy-trends-for-2021/.

[28] Ted Galen Carpenter, *Many predicted NATO expansion would lead to war. Those warnings were ignored.* **The Guardian**. Monday, 28 Feb 2022 19:00 GMT.

[29] John Mearsheimer on why the West is principally responsible for the Ukrainian crisis, ECONOMIST, March 11, 2022. economist.com.

irreconcilable value conflicts within and among societies."[30]True universality of studies of this nature may not be possible. Security studies are highly "dependent on spatio-temporal conditions."[31]

However, for as long as nation-states continue to play the roles that they uniquely perform in the world community, the generalizations of this study, in the author's opinion, can remain as foundations and background of Philippine experience in the international scene.

Despite the foregoing limitations, development goals can be achieved with the help of this study. As the results show, they can become inputs to the shaping of Philippine institutions, correcting unacceptable practices and inappropriate behavior that are not consistent with national goals.

The author hopes that the readers will gain expertise, if not mastery, of the classical realist theory and a working knowledge of at least one of its variants, especially the concept of neorealism. They will become, at the very least, familiar with Philippine practice. This type of research, assessing the variation between theory and practice of realism in the country, is not very common in the Philippines. Hence, the applicability of this field of study will be of interest to foreign policy makers, experts, and practitioners of the discipline.

---

[30] Felix Rosch and Richard Ned Lebow, *A Contemporary Perspective on Realism*, E-International Relations, p.1. https://www.e-ir.info/2018/02/17/a-contemporary-perspective-on-realism.

[31] Ibid. p. 3.

Chapter 2: Realism and Related Literature at Various Eras

Overview of Other Work in Different Times

Chinese Model of International Relations in Asia

"The starting point for understanding the problem is to recognize that China is not just another nation-state in the family of nations. China is a civilization pretending to be a state."[32]

China is a country with ancient history and civilization. It has dealt with foreign powers throughout its national narrative. It has its own ethnocentric view of its nationhood, as in believing that it is the Middle Kingdom. It is known to have invented gunpowder. Its influence is felt everywhere in the world. Yet, there was a time in its history when it was subjugated by major powers, mainly the Western ones, but an Asian power, Japan, also had a hand in invading Chinese territory.

China has its own system in every human endeavor. Hierarchy is important. Reverence for elders and the past are part of its culture. China was on its own footing until its system was threatened by the different order of nations created by the Peace of Westphalia in 1648, which granted European powers independence and boundaries. Thus, diplomacy was born in Europe. The pride of the Chinese is manifested by the fact that "the Chinese experience is uniquely challenging, as it evolved

---

[32] Lucien Pye, "China: Erratic State, Frustrated Society," **Foreign Affairs**, 69:4 (Fall 1990), p. 58, quoted in Yongjin Zhang, *System, empire and state in Chinese international relations, **Review of International Studies**, Vol. 27, Special Issue: Empires, Systems and States: Great Transformations in International Politics* (December 2001), p. 63, Published by: Cambridge University Press. Stable URL: https://www.jstor.org/stable/45299504.

entirely independently of European influence until modern times."[33]

It is amazing to note that, because of the age of Chinese civilization, there are traces of the elements of the realist theory in its ancient, early versions. The systemic configuration of the Chinese states "faced the classical dilemma of any decentralized international system, that of anarchy."[34] Let us recall that anarchy is one of the basic characteristics of the international system operating under the realist doctrine. The parallel developments in China and in Europe may be viewed as components of the same human experience, albeit in different places, unrelated culturally and geopolitically. Yongjin Zhang has argued "that elements of constitutional principles and some basic institutional practices that are said to have characterized the modern European international system were already present and functioning in the system of states in Ancient China."[35] It can further be posited, therefore, that perhaps China is ahead of Europe in international relations theory. The Chinese system can even be alleged to be superior because the Chinese states were identified as having had a common culture that transcended their provincial boundaries.

The Chinese system of states appeared to have had international law among themselves. The concepts of conflict resolution and management were not alien to them. In their relations with each other, present day principles like "sovereignty, diplomacy, the balance of power and rituals"[36] were practiced. Balance of power is one of the key ideas of realism. The expression of external sovereignty may not have been verbalized as such. However, the exercise of sovereignty is apparent as "Ancient

---

[33] Yongjin Zhang, *System, empire and state in Chinese international relations,* Ibid., p. 44.

[34] *Ibid.* p. 46.

[35] *Ibid.*

[36] *Ibid.* p. 47.

Chinese states, for example, monopolized the right to declare war against each other."[37] It can be argued that perhaps all of the modern practices today that were derived from European diplomacy existed abundantly among Chinese states. Such activities as treaty-making, cession of territory to gain peace, control of passage through territory by foreign diplomats, right to offer political asylum and extradition[38] were part of the activities of statehood.

For purposes of this analysis, the most important postulate is balance of power. It was "the most vital institution that sustained the existence of the Ancient Chinese system of states. It is in fact difficult to contest the argument that it is the collapse of the balance of power that led to the establishment of the first Chinese empire in 221 BC by the Qin state."[39] The role of balance of power has been chronicled in Chinese classics. The rise and fall of states tell stories of how they played "the balance of power game of survival, protection, conservation and expansion."[40]

While the realist principles of managing foreign relations already existed in Ancient China on the basis of its anarchic nature, there actually was a system that preceded anarchy. We will mention it here, for historical purposes, to point out how this concept differed from the European order, which was also derived from anarchy. This is "the Chinese belief in a hierarchical cosmic order within which every being was assigned a proper place."[41] Hierarchy gave way to anarchy, the enduring characteristic of the international system to this very day.

---

[37] *Ibid.* pp. 47-48.

[38] *Ibid.* p. 48.

[39] *Ibid.* p. 49.

[40] *Ibid.*

[41] *Ibid.*

One interesting highlight of Chinese diplomatic practice was the prevalence of rituals. What is now known as protocol or ceremonials were widespread. These rituals dictate behavior in particular circumstances. "Even in such a narrow conceptualization, strict observation was required. A large number of diplomatic representations were obliged and reciprocated among states simply for the purpose of observing rituals, for example, of the assumption of the throne and the death of a ruler."[42]

The simultaneous existence of European world order and the Chinese sub-global international system could not have been expected to last indefinitely. One of them had to give way to the other. The Westphalian order took over in the second half of the nineteenth century.[43] This is a major change in the international scene.

What caused the imperial Chinese system to collapse? Zhang posits two reasons: "One was the loss of tributary states along the periphery of the Empire. The expansion of Europe in the form of British, French and Russian imperialism, and later Japanese imperial expansion, reached the peripheral areas of the Middle Kingdom from the 1870s onwards."[44] The second reason is the more painful of the two. It was the ultimate nail in the coffin. It was "the crumbling of the entire pack of imperial institutions and the final collapse of the imperial polity *per se*, that rendered imperial collapse irrevocable."[45] The result was the humbling of China, in effect bowing to European superiority in international practice. Indeed, China thus just became one country among many, under the leadership of the European powers. It was not a pleasant experience for the Middle Kingdom. Zhang says, "Acknowledging

---

[42] *Ibid.* p. 50.

[43] *Ibid.* p. 52.

[44] *Ibid.* p. 60.

[45] *Ibid.*

16

sovereignty and equality as the most fundamental principle in China's international relations amounted to admitting the irrelevance of basic assumptions of the Sino-centric view of the world."[46] Chinese political philosophy is based on Confucianism, essentially the constitutive principle of the Chinese world order and informs systemic norms of procedural justice.[47] The fall of the tribute system, among others, which was the building block of peacetime diplomacy, ensured that a new system sought to fill in that vacuum (see the comparison of the tribute system with other fundamental institutions of societies, Table 1).

It was the end of an era and the beginning of another.

---

[46] *Ibid.*
[47] *Ibid.* p. 56.

Table 1. Constitutional structures and fundamental institutions of international societies: a comparison[48]

Societies of States	Ancient Greece	Imperial China	Modern Society of States
Constitutional Structures			
1. Moral Purpose of State	Cultivation of *bios politikos*	Promoting cosmic and social harmony	Augmentation of individuals' purposes and potentialities
2. Organizing principle of sovereignty	Democratic sovereignty	Sovereign hierarchy (civilizational)	Liberal sovereignty
3. Systemic norm of procedural justice	Discursive justice	Ritual justice	Legislative justice
Fundamental Institutions	Interstate Arbitration	Tribute System	Contractual International law multilateralism

---

[48] This table is found in Yongjin Zhang, *Ibid.* p. 57. It quotes the source of the table as Christian Reus-Smit, **The Moral Purpose of the State: Culture, Social Identity, and Institutional Rationality in International Relations** (Princeton, NJ: Princeton University Press, 1999, p. 7).

Chinese Foreign Policy Making in the Modern Era.

Let us fast forward to modern times. China has become one of the major political, economic, and military powers of the world. Conscious of its stature, it is always protecting its national interest. Using the neoclassical realist theory of international relations, Yufan Hao and Ying Hou made a comparative study[49] of the foreign policy formulation process between China and the United States. With the following diagram (Fig. 1)[50] of the neoclassical realist approach as a framework, the two authors made some interesting findings. The authors say that "China's regime is more conducive to the continuity of foreign policy."[51]

Fig. 1. The Neoclassical Realist Approach

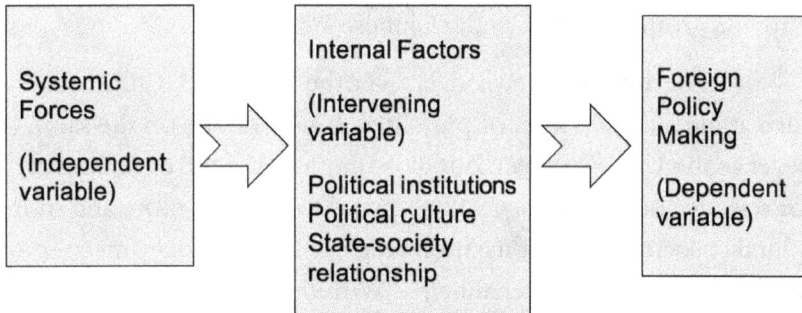

| Systemic Forces (Independent variable) | Internal Factors (Intervening variable) Political institutions Political culture State-society relationship | Foreign Policy Making (Dependent variable) |

The Neoclassical Realist Approach. The authors state that, "In fact, every time an American president with a different party background from his predecessor comes into power, Chinese

---

[49] Yufan Hao and Ying Hou, *Chinese Foreign Policy Making: A Comparative Perspective*, **Public Administration Review**, Dec., 2009, Vol. 69, Supplement to Volume 69: Comparative Chinese/American Public Administration (Dec., 2009), pp. S136-S141. Published by: Wiley on behalf of the American Society for Public Administration https://www.jstor.org/stable/40469084.

[50] Fig. 1 is adopted from Yufan Hao and Ying Hou, *Ibid*. p. S137.

[51] *Ibid*. p. S137.

leaders face the problem of what foreign policy the new president will adopt and how best to deal with this new president."[52] How true indeed! It is not only the Chinese who have this problem. All leaders the world over have to figure out what is in the mind of the new U.S. president in dealing with other countries.

Another point of the above-mentioned comparative work is that "unlike China's unitary legislature and administration, the separation of powers in the United States provides an opportunity for the executive and Congress to compete in foreign policy making."[53] This would cause undue delay in the process and the long, drawn-out hearings provide foreign powers with a ringside seat on the debates that sometimes reveal the inner workings of the American bureaucracy.

The long tenure of Chinese government officials, compared with the short-term electoral cycles in the US, appears to provide greater stability to the Chinese system.

The author believes that, over the long-term, China will also attain a certain level of pluralism. It may not be on the same level as the United States "But globalization and the development of science and technology will continue to enable a more and more pluralist society to exert itself on Chinese foreign policy making through every possible channel."[54] With this assumption, there is every hope that the Chinese and the American systems will approach a middle ground. The authors are not so fast to assume that China's progress in this regard will approximate that of the West's. "Just as the neoclassical realist analytical framework shows, China's internal factors—its peculiar culture, historical experience and national characteristics—will make Chinese development as

---

[52] *Ibid.* p. S138.
[53] *Ibid.*
[54] *Ibid.* p. S140.

well as foreign policy making unique with 'Chinese characteristics.'"[55]

Table 2. Evolution into a Uniform Practice of Diplomacy[56]

Estimated period	China	Europe
1500 1600 1700 1800	Hierarchical Tribute system	Individualistic Anarchical
Mid 1800s 1900 2000	Westphalian system	Westphalian system

Table 2 depicts the evolution of the Chinese system of international relations into the Westphalian system recognizing the sovereignty of individual states, their boundaries and their distinctive citizens. China was compelled by Western powers to adapt to the latter's ways.

An emerging question is whether China, with its tremendous economic growth, will begin to openly challenge the lone hegemon, the unipolar power, the United States. There are indications that China's neighbors feel threatened, causing

---

[55] *Ibid.* p. S141.

[56] Author's own interpretation of the relevant periods covered by Yufan Hao and Ying Hou, *supra.*

Southeast Asian countries to be wary and improve their defenses, thus causing an arms race in the region.

A neoclassical realist approach[57] to determine China's aggressiveness says that, in order to analyze this issue, the role of Chinese leadership and relevant domestic factors must be part of the inputs as intervening variables between systemic factors and the resulting foreign policy. This is shown by Figure 1, *supra*.

Sorensen concludes that there are conflicting demands on Chinese leadership: both internally and externally. "China's external behavior is often seen by the international community as being irresponsible while also viewed by the Chinese population as too weak."[58] This push and pull is further complicated by nationalist elements in the People's Liberation Army. "Consequently, China's external behavior, rather than reflecting greater self-confidence and aggressiveness, actually reveals an internally conflicted, inward-looking, and reactive China."[59]

Overview of Spanish Foreign Relations.

Spain was one of the dominant colonizing powers well into the 16th century until the end of the 19th century. Under the Treaty of Tordesillas, Spain and Portugal divided the world into their respective spheres of influence. The year 1492 was significant to Spain because that was the year that it expelled the last Muslim

---

[57] Camilla T. N. Sorensen, **Is China Becoming More Aggressive? A Neoclassical Realist Analysis**, *Asian Perspective*, July-Sept 2013, Vol. 37, No. 3 (July-Sept. 2013), pp. 363-385. https://www.jstor.org/stable/42704834.
[58] *Ibid*. p. 380.
[59] *Ibid*.

ruler of Granada. It was also the year that Christopher Columbus sailed through the Atlantic Ocean to the Americas.[60]

It is interesting to note that the impetus for Spanish (and other powers at that time) exploration of countries like the Philippines was the search for spices, tea, silk and other exotic Oriental products. In general, the search for wealth, like silver, was part of the plan to ship these precious metals to pay Spain's debts, finance its armies and construct churches. Silver from the Americas was transported to the Philippines as well.[61]

The Treaty of Westphalia had a direct negative effect on Spanish interest. As a result of the end of the Thirty Years' War, the Netherlands gained independence from Spain. It was one of the early losses of the colonies of Spain, and a vital territory indeed as it is in Europe. Unlike China, Spain was within the orbit of the Treaty of Westphalia early on. Signed in 1648, it remains to this day as the basis of the European-based international order.

Spanish vital interests and national power rested on its identity as a Catholic country. Along with the *conquistadores* came the bearers of the cross, the friars. The mode of subjugation of natives in the colonies was a clear example of the unity of church and state.

The Catholicism of Spain was perhaps its Achilles heel. There is a theory that the expulsion of the Jews in 1492[62] removed

---

[60] Ringmar, Erik, *History of International Relations: A Non-European Perspective*, New Edition [online], Cambridge: Open Book Publishers, 2019, **Chapter 8, European Expansion,** http://books.openedition.org/obp/9111, ISBN: 9781783740246.

[61] *Ibid.*

[62] Henry Kamen, *The Mediterranean and the Expulsion of Spanish Jews in 1492*, **Past and Present**, May 1988, No. 119 (May, 1988), pp. 30-55, Oxford University Press on behalf of The Past and Present Society, https://www.jstor.org/stable/651019.

a portion of the Spanish population that had the potential of creating a strong middle class. It is believed that the industry and business acumen of the Jews were part of the national wealth of the country. If the latter suggestion is true, the expulsion of the Jewish population backfired against Spain. On this premise, the realist presumption that Spain was working to achieve its best national interests, using its raw power in the process of expelling the Jews, did not achieve the desired result.

Spain rapidly declined as a world power when it lost its colonies, starting from those in the Americas. In 1898, it lost the Philippines to the United States,[63] where the latter paid the Spaniards 20 million dollars under the terms of the Treaty of Paris. Other than close cultural ties, the relationship between the Philippines and Spain slowly faded into the background. This was all the more apparent after the Spanish Civil War and during the Second World War, as Spain, under the dictatorship of Francisco Franco, allied itself with fascist regimes.[64]

Today, Spain is a member of both the EU and NATO, truly a beneficiary of Western European integration. It has one of the best economies in the world and it is clear-eyed about its national interest and power.

---

[63] Linda Robinson, Patrick B. Johnston and Gillian S. Oak, *U. S.-Philippine Relations in Historical Perspective*, **U. S. Special Operations Forces in the Philippines, 2001-2014**. p. 9. RAND Corporation, https://www.jstor.org/stable/10.7249/j.ctt1cd0md9.9.

[64] H. L., Spain: Foreign Relations and Policy since 1940, *Bulletin of International News*, Nov. 14, 1942, Vol. 19, No. 23 (Nov. 14, 1942), pp. 1013-1018, Royal Institute of International Affairs, https://www.jstor.org/stable/25643333.

A Snapshot of British International Affairs

Great Britain's foreign policy adheres to realism. It was a master of the balance of power system that existed in Europe, especially during the Napoleonic Wars. In the Second World War, it was Hitler's invasion of British ally Poland that ignited the war. With the attack on Poland, Britain had to declare war on Germany pursuant to an alliance commitment.

The Royal Navy was second to none, until the arrival of the United States in the war. After the defeat of Germany and Japan, Great Britain and the U.S. forged one of the closest alliances in history. To this day, that alliance is very much alive and in full swing.

In addition to the schools of thought on foreign policy analysis earlier mentioned, there is a concept in the United Kingdom called the English School. This international relations theory says that, in addition to anarchy in the system, there is also such a thing as "a society of states." The English School is sometimes known as liberal realism. The society of states engenders cooperation, while realism, *per se,* connotes unceasing struggle and competition.

The English School, according to Barry Buzan, is based on the concept of "the international society." He claims that he will "use the logic of structural realism to show how international society can emerge as a natural product of the logic of anarchy."[65]

---

[65] Barry Buzan, *From International System to International Society: Structural Realism and Regime Theory Meet the English School,* **International Organization,** Summer, 1993, Vol. 47, No. 3 (Summer, 1993) p. 327.

Foreign Affairs of the United States.

The United States had experienced what many countries had to go through in their history. It suffered under British colonial policies, it had its own horrific civil war, and it had to balance its relations between European powers and the newly-emerging states of the American continent. The young republic realized that it had to stand on its own in protecting its national interests.

Dexter Perkins, a historian and an expert on the Monroe Doctrine, set out the moorings of American foreign policy in its growing-up years. His premise is that "the foreign policy of a democracy must and will differ from that of a totalitarian state."[66] At the outset, Perkins laid the ground rules: the United States is a democratic state and it is diametrically opposed to some of the regimes in Europe, which was then under a cloud of war. Immediately, there was distancing and differentiation from the fascist and dictatorial states engulfing the continent.

As a young nation, the United States had to contend with the politics of European powers. In its early years, having itself just gained independence, it tried its best to feel its way into the international community. It was not easy because all of them, naturally, had different and even opposing, national interests. And, it is interesting to note that the Europeans have never lost sight of the American continent. Even if the United States was lost to Britain, and became an independent country, there were other territories, or even states, that were ripe for exploitation by the Old World powers.

---

[66] Dexter Perkins, *Fundamental Principles of American Foreign Policy*, **The Annals of the American Academy of Political and Social Science,** Nov., 1941, Vol. 218, Public Policy in a World at War (Nov., 1941), p. 9.

Table 3. Mass-based Policy Formulation Differences[67]

State	Character of Public Opinion
Democratic	The opinion of the mass is not manufactured but grows; the process is no doubt a complex and slow one; it is no doubt conditioned by many private interests and by many selfish views; but it will never be free from the broad general conceptions of both a political and a moral character. Maintains a higher standard of good faith.
Totalitarian	Perfect shameless cynicism, complete faithlessness, brutal opportunism; Manufactures by its machinery of propaganda and organized mendacity the public opinion necessary for the carrying out of its policies.

It was around this point that the United States, pursuant to its policy of avoiding "entangling alliances," issued what became known as the Monroe Doctrine. President James Monroe, the fifth president of the United States, essentially prohibited European powers from involving themselves in the internal affairs of the

---

[67] *Ibid.* This table was generated from the discussion by Perkins on page 9 of his article explaining the differences in mass-based approach between democratic and totalitarian states.

various countries in the Americas. The Monroe Doctrine appears to still be in effect up to this very day, as the Cuban Missile Crisis and the Nicaragua-Soviet cooperation have shown. In both instances, U. S. assertion of the Monroe Doctrine, albeit unstated and under the shadow of the Cold War, resulted in Soviet withdrawal from the hemisphere.

Along with the Monroe Doctrine came another bedrock principle in American foreign policy: neutrality in European affairs. This stance was broken by the First and Second World Wars and the post-war commitment to NATO.

The post-war scenario resulted in bipolarity in international affairs. This period was characterized by the tug-of-war in all spheres of human endeavor between the United States and the Soviet Union. As a result of their ideological struggle, there appeared volumes of literature on international security and international relations theory. One of the respected early cold war pundits was George F. Kennan, a former U. S. diplomat, who was stationed in Moscow. He was the author of the policy of containment. Under this policy, the Soviet Union's sphere of influence was to be limited within its borders. Writing as "Mr. X," Kennan was a "supreme realist"[68] and, later in life, advocated positions that were contradictory to American government policies, like a unified but neutral Germany.[69]

Kennan foreshadowed today's events. In an interview in 1989, he referred to the mission direction of U. S. and NATO forces in these words: "the tradition of our armed services, and of all the NATO armed services, has been to take into account only the capabilities of our conceivable foe, but not his interests or his

---

[68] Christian Caryl, *The Enigma of Mr. X*, a review of **The Kennan Diaries** by George F. Kennan and Frank Costigliola, <u>The National Interest</u>, No. 130, THE GOP'S BALANCING ACT (March/April 2014), p. 80.
[69] *Ibid.*

intentions or his incentives. This is to my mind today unrealistic."[70] It is as if the latter statement is being uttered at this moment. Many analysts say that Putin ordered the invasion of Ukraine because his interests, sentiments, and need for security, were brushed aside by the Western powers led by the United States and NATO.

The U. S. suffered a debacle in Vietnam. But in that era, another realist, Henry Kissinger, helped Nixon open China to the world. Then came September 11, 2001, and the subsequent invasion of Afghanistan. The chaotic withdrawal from Kabul two decades later showed lack of American resolve to stay, eliminate terrorism, and plant democracy in that country. Questions linger on American commitment to its supposed partners in democratization and counterterrorism.

Iraq was also invaded principally for the U. S. to impose regime change. The U. S. succeeded in changing the government but, to this day, American physical presence is still felt in Iraq. In all these years, the U. S. benefited from the brains of realist theoreticians in its bureaucracy and academic institutions. Worth mentioning as well are the *neocons*, or neoconservatives, who were either in the State Department or the Pentagon. The *neocons* appear to be highly concerned with power, intervention abroad, and of course, the national interest of the United States in their policy recommendations. It would not be surprising to find that realists and *neocons* coincide in their thinking on international affairs. *Neocons*, however, are not merely concerned with foreign relations, but they are immersed in domestic policy as well, as in minimal government and a free-market operation of the economy.

---

[70] George F. Kennan and Arms Control Today, *George F. Kennan On the Crisis Within the Soviet System*, **Arms Control Today,** September 1989, 19:7 (September 1989), pp. 3-9.

Greek Legacy and Roman Influence

Any approach to an issue using history as a tool would not be complete without resorting to traces of Greek and Roman influence. Greece and Rome gave the world systems and philosophies of government whose footprints are still visible in our day. In his article on Montesquieu's views on ancient Greek foreign relations, Andrea Radasanu says, "Montesquieu weaves a modernization tale from primitive ancient Greece to modern commercial states, all to teach the reader to overcome any lingering attachment to glory and to adopt the rational standards of national interest and self-preservation."[71] (underscoring supplied). These two standards are basic guidelines underpinning realist principles. In ancient Greece, there were incessant wars. Conflicts among nations and ethnic groups were settled in the battlefield. Yet, the competition in arms slowly had to give way to development, no matter how slow. "Serious study of the Greek tale discloses that its rhetorical purpose is to prompt readers to overcome whatever lingering appreciation they have for the heroic ideal and to shift the narrative of international relations toward considerations of power and prosperity for the sake of peace."[72]

According to Radasanu, Montesquieu regards the Greeks more favorably than the Romans in their civilizational influence.[73] Montesquieu also said that "Greek imperial ambitions (particularly those of Alexander) were somewhat progressive and prefigured

---

[71] Andrea Radasanu, *Montesquieu on Ancient Greek Foreign Relations: Toward National Self-Interest and International Peace*, **Political Research Quarterly,** MARCH 2013, Vol. 66, No. 1 (MARCH 2013), p. 3 (Abstract). Sage Publications, Inc. on behalf of the University of Utah https://www.jstor.org/stable/23563585.

[72] *Ibid.* p. 4.

[73] *Ibid.*

modern commerce and, so, laid the groundwork for peace among nations."[74]

The martial spirit of Greece and Rome did much to solidify their hold on foreign territories that fell under their yoke. Military control was an absolute necessity. "Spartan *esprit de corps* lent itself to security, and Roman grandeur or *grandezza* produced brilliant if ultimately vulnerable empires."[75]

Radasanu says that a realist outlook has played a major role in Montesquieu's analysis of the Greek story. According to Radasanu, "Montesquieu's hopes for peace are firmly rooted in his realism, but his realism is informed by his study of human nature and the changes and distortions that affect this nature."[76] The role of human nature in realist thinking has a big part in Morgenthau's modern realist theory of international relations.

Development of Realism as a Theory.

Before the prominence of realist ideas, one prevalent view was the inherent goodness of human beings. It was thought that relations between them are underpinned by neighborly ties, with the undercurrent of kindness and friendship. Religious leaders and writers wrote extensively on this aspect of human relations. Two thinkers in particular, namely, Jean Jacques Rousseau and Douglas McGregor, spoke of the "optimistic image of innate human nature."[77] Rousseau specifically "believed that in the state of nature

---

[74] *Ibid.*

[75] *Ibid.*

[76] *Ibid.* p. 13.

[77] William G. Scott and David K. Hart, *The Moral Nature of Man in Organizations: A Comparative Analysis*, **The Academy of Management Journal**, Jun., 1971, Vol. 14, No. 2 (Jun., 1971), p. 250. Academy of Management. https://www.jstor.org/stable/255310.

man was born with an innate predisposition toward self-preservation, mitigated by a compassion for all other men."[78]

Elements of the realist theory of international relations date back thousands of years. One of the earliest proponents of the discipline was Thucydides. He chronicled his work in a book entitled *"History of the Peloponnesian War."*[79] In his writings, Thucydides highlighted the primacy of power and national interest but added political and cultural differences[80] in the analytical framework. Many scholars have recognized Thucydides as a pioneer in realism. Paul Viotti, Mark Kauppi, Michael Doyle, Kenneth Waltz, Robert Keohane and Joseph Nye, among others, acclaim the realist leanings, if not excellence, of Thucydides.[81]

A writer named Laurie M. Johnson Bagby puts the ideas of Thucydides within the context of his time. It is here that Thucydides can be claimed to have written his thoughts that distance themselves from realist theory as known today. Thucydides departed from realist theory in the following methodologies: "(1) emphasizing the importance of what we might call 'national character,' (2) highlighting the influence of the moral and intellectual character of individual leaders, (3) showing the importance of political rhetoric for action and treating what we call realism as another argument in political rhetoric, not a theory that Thucydides thinks describes the whole truth about political things, and (4) showing that for him, moral judgments form an integral part of political analysis."[82]

---

[78] *Ibid.*

[79] Early fourth century BC.

[80] Laurie M. Johnson Bagby, *The Use and Abuse of Thucydides in International Relations*, **International Organization**, Winter, 1994, Vol. 48, No. 1 (Winter, 1994), pp. 132-133. The MIT Press. https://www.jstor.org/stable/2706917.

[81] *Ibid.* p. 131.

[82] *Ibid.* p. 133.

While experts in international relations concede that Thucydides was one of the standard bearers of realism, there are authors like Bagby who view him in a more comprehensive way, that is, adding the qualities of leadership and their moral character as part of the critical factors that determine the success (or failure) of a nation.

Hundreds of years later, another author, Machiavelli, wrote that "the end justifies the means."[83] According to Machiavelli, the supremacy of the state trumps everything, including morality.[84] Machiavelli differed from his predecessors and was much closer to Morgenthau. The classical or medieval thinkers who were ahead of Machiavelli "took their political bearings from transcendentally valid or divinely sanctioned conceptions of justice."[85] In fact, Machiavelli "oriented himself as the 'effectual truth' of politics: how the world actually 'is' rather than how it 'ought' to be."[86] The latter phrase was to be echoed and re-echoed many years later by Morgenthau. Among Machiavelli's famous dictums was, "it is safer for a prince to be feared rather than loved."[87] The reason for this level of comfort for a prince is that "subjects love at their own pleasure while they fear at the pleasure of a prince."[88]

A subsequent scholar, Hobbes (*Leviathan*[89]), described human life as "nasty, brutish and short," characterized by never-ending competition. This analysis is applicable as well to that of the state, to whom individuals belong, in its relationship with other

---

[83] *The Prince,* 1532.
[84] *Ibid.*
[85] John P. McCormick, ***Social Research,*** Vol. 81, No. **1, MACHIAVELLI'S THE PRINCE 500 YEARS LATER** (SPRING 2014], The Johns Hopkins University Press. p. xxiv.
[86] *Ibid.*
[87] *Ibid.*
[88] *Ibid.*
[89] Written in 1651.

states. Hobbes' grim view of human character fits right into the realist lens. "The state of nature as Hobbes describes it, where trust is nonexistent, one is forced to act as if one's gain were always another's loss, even though, as Hobbes points out, the collective outcome of such action is loss for almost everyone."[90] One wonders if the "loss for almost everyone" is applicable to the occupation of another country's territory.

One school of thought that, among others, upholds the inherent goodness of humans is *idealism*, discussed earlier. Realism is far removed from idealism, if not its opposite. Thucydides, Machiavelli, and Hobbes say that humans fend for themselves at the expense of others. E.H. Carr attacked idealism.[91] He wrote that human nature in reality centers on self-interest. Carr differed from those who believed in utopian solutions like "Arnold Toynbee, Norman Angell, and Alfred Zimmern."[92] Carr's language reminds readers of Morgenthau's tone when Carr said that, "students and theorists of international politics tend to ignore what is and put forward utopian schemes for what should be."[93] He highlights the work of diplomats and bureaucrats with the claim that "officials who have to live with the realities of diplomacy tend to be much more realistic."[94] Consistent with the writings of Morgenthau, Carr entertained "a conviction that the international system rested on power, and only a realistic appreciation of relative power and the

---

[90] James H. Read, *Polity*, Summer, 1991, Vol. 23, No. 4 (Summer, 1991), p. 506. The University of Chicago Press on behalf of the Northeastern Political Science Association.

[91] *The Twenty Years' Crisis, 1919-1939: An Introduction to the Study of International Relations,* London: Macmillan and Company, Limited, 1939.

[92] J. D. B. Miller, *E.H. Carr: The Realist's Realist*, Reviewing the work: **The Twenty Years' Crisis, 1919-1939: An Introduction to the Study of International Relations** by E. H. Carr, supra. p. 65. The National Interest, Fall 1991, No. 25 (Fall 1991), Center for the National Interest.

[93] *Ibid.*

[94] *Ibid.*

readiness of national leaders to use it could provide a sound basis for policy."[95]

Morgenthau espoused slightly different but refined views, and his work was more scientific in that it sought to predict the behavior of nations on the basis of power. He came up with basic principles; six of them (*supra*) were to govern international relations. It is around these six basic principles that this study aspires to focus on. They are so fundamental that many students and university graduates may at least have heard of them, if not studied them in political science, history, or international relations courses. The articulation of a country's placement in the international order has to have some measurement, a gauge with which it is able to project itself. Thus, "The concept of the national interest defined in terms of power was, of course, central to Morgenthau's thinking."[96] All nations ought to take care of their respective national interests if they can be viewed as significant participants in the world stage. Morgenthau "used the national-interest concept as a yardstick, and by applying this yardstick to all foreign policies made it possible to make a meaningful comparison between, for example, the policies of Pericles and of F.D.R."[97] It is like science in its use of mathematics as a measuring tool. One may know the speed of an object, but such a phenomenon should be capable of being gauged in terms of its direction and exacting nature, so that the concept can provide accuracy, certainty, if possible, and a measure of predictability.

Morgenthau's foundational principles were modified to be more scientific by Kenneth Waltz in his book *Theory of International Politics*, published in the late 1970s. Waltz proclaimed that the

---

[95] *Ibid.* p. 67.
[96] David Fromkin, *Remembering Hans Morgenthau*, **World Policy Journal**, Fall 1993, Vol. 10, No. 3 (Fall, 1993) Duke University Press, p. 84.
[97] *Ibid.*

"international system" and its anarchical nature, keep order for every member of that system. Waltz's major failure was his inability to take account of the fall of the Soviet Union. In a rebuttal against the so-called irrelevance of structural realism after the cold war, Waltz said that the Cold War "is firmly rooted in the structure of postwar international politics and will last as long as that structure endures."[98] Waltz defended his theory from detractors because, according to him, the Cold War ensued only after "changes in the system," not "changes of the system."[99] Thus, Waltz has not surrendered in the debate on whether realism in international politics has been rendered obsolete.

In an article, Stephen M. Walt concluded that, "If you think like a realist, in short, you are more likely to act with a degree of prudence, and you'll be less likely to see opponents as purely evil (or see one's own country as wholly virtuous) and less likely to embark on open-ended moral crusades. Ironically, if more people thought like realists, the prospects for peace would go up."[100] Walt was alluding to the primacy of states and the power that they possess.

The application of realist principles cannot fully predict future events, as hinted earlier. Predictions may succeed and they may fail. And, "no matter how well designed the structure of political institutions, power will always be the ultimate arbiter of outcomes in international politics."[101]

---

[98] Kenneth Waltz, *"Structural Realism after the Cold War,"* **International Security**, Summer, 2000, Vol. 25, No. 1, p. 39, quoting himself in a previous work entitled *"The Origins of War in Neorealist Theory,"* **Journal of Interdisciplinary History**, Vol. 18, No. 4, (Spring 1988), p. 628.
[99] *Ibid.* p. 5.
[100] Stephen Walt, *The World Wants You to Think Like a Realist*, **Foreign Policy**, p. 6, May 30, 2018.
[101] Samuel Barkin, *Realism, Prediction and Foreign Policy*, **Foreign Policy Analysis**, July 2009, Vol. 5, No. 3, pp. 233-246.

Felix Rosch and Richard Ned Lebow say that realism is here to stay, especially in its classical form. Realism, according to them, "is therefore far from being ready for the dustbin of the history of International Relations (IR) theory–as some critics suggest. It can serve as a stepping stone to question some of the common assumptions held in the discipline, propose solutions to some of the contemporary problems in international relations and show us how we can create more inclusive societies."[102]

Arguably one of the best diplomats of the U.S. after the post-World War II era was Ambassador George Kennan. He served in Eastern Europe and was Ambassador to the Soviet Union. He was an advocate of the "containment policy" that upheld U.S. national interest vis-a-vis the USSR. He wrote what is called the "Long Telegram" explaining Soviet motivations and background of their behavior against U.S. interests and Western nations in general. As a realist, he looked far beyond his tour of duty in Moscow. Wrapping up his lengthy report, he said, "In summary, we have here a political force committed fanatically to the belief that with the US there can be no permanent *modus vivendi*, that it is desirable and necessary that the internal harmony of our society be disrupted, our traditional way of life be destroyed, the international authority of our state be broken, if Soviet power is to be secure."[103] A classic case of a real security dilemma.

---

[102] Felix Rosch and Richard Ned Lebow, *A Contemporary Perspective on Realism*, https://www.e-ir.info/2018/02/17/a-contemporary-perspective-on-realism/ Feb. 17, 2018.

[103] George Kennan's *Long Telegram*, February 22, 1946, History and Public Policy Program Digital Archive, National Archives and Records Administration, Department of State Records (Record Group 59), Central Decimal File, 1945-1949, 861.00/2-2246; reprinted in US Department of State, ed.,Foreign Relations of the United States, 1946, Volume VI, Eastern Europe; The Soviet Union (Washington, DC: United States Government Printing Office, 1969) p. 14 of declassified original document. https://digitalarchive.wilsoncenter.org/document/116178.

There is little work done on the application of realism to the Philippine situation. One article asks whether a state can entrust its security to multilateral security institutions. This question arises out of the weakness of the Philippine security apparatus. The writer, Leszek Buszynski, asserted that, "The Philippine move to involve the U. S. in a balance of power was a realist response that demonstrated that it would be premature for small or medium powers in similar situations to base their security on institutionalism alone."[104]

Prof. Buszynski's piece highlights the need for the Philippines to undertake self-reliance defense programs that will reduce its dependence on other states for its own security. The Philippines cannot be fully independent as a state if it continues to ask other countries to bear the burden of its territorial protection, which is one of the sovereign duties of an independent nation.

According to Buszynski, the Philippines has not been successful in enlisting the support of ASEAN in its security requirements vis-a-vis China. Instead, it has gravitated towards the United States, despite the eviction of its troops in 1991. Thus, the Philippine failure in ASEAN, and in the ARF, left it with no option but to take the realist approach by invoking its old alliance with the U.S.[105]

Michael Magcamit, in his article, concludes that President Duterte's change of direction towards China in his policy pronouncements is a paradigm shift to ensure survival.[106]

---

[104] Leszek Buszynski, *Realism, Institutionalism, and Philippine Security,* **Asian Survey**, 42:3, p. 500.

[105] Ibid., pp. 491-496.

[106] Michael Magcamit, *"The Duterte method: A neoclassical realist guide to understanding a small power's foreign policy and strategic behaviour in the Asia Pacific,"* **Asian Journal of Comparative Politics**, October, 2019.

All the studies we have seen point to the state as the center of relationships in the international system. The aim of this study is to show the experience of the Philippines within the theoretical framework of realism. The Buszynski article is a step in that direction. Magcamit made the study from the point of view of the president, the chief architect of foreign policy and the commander-in-chief of the armed forces.

The early studies on realism, e. g., by Morgenthau et al., provide theoretical approaches, although their empirical evidence is rooted in history. This research will focus on the Philippines and intends to convey a practical example of state experience.

While the author of this study will heavily rely on Morgenthau, other scholars who worked hard in the discipline of international relations might be worth mentioning here. Robert Keohane believed that international institutions must cooperate with each other. Robert Cox advocated social emancipation. He disagreed with the ideas of control and domination that neorealists espoused. One author who subscribes to constructivism is Alexander Wendt. His theory is that while constructivism can account for change in world politics, neorealism cannot. He says that, to reconcile these two scholarly approaches, those in the profession must concentrate in the development and formation of interests and identities.

There is also the international relations English School that advocates a peaceful environment through reciprocity and cooperation. Society, according to the English School, is capable of engendering mutual respect among human beings.

Theoretical/Conceptual Framework

The foundation laid down by Morgenthau is the framework that will be used in this work. The author's concept is that the Philippines has not fulfilled Morgenthau's thesis that the quest for power is inherent in the survival of all states. A contemporary reading by the author of the conduct in the Philippines' foreign relations indicates that the country is deficient in its quest for power. There have been developments in the recent past that the Philippines is seeking modern weapons[107] to hold other countries at bay in the South China Sea. It is a positive development, but it is a long way from self-sufficiency in its defense needs. A self-reliant defense posture can only be achieved through industrial development and scientific advancement. Economic progress is an indispensable element of national power.

This study will propose to broaden the approach to include foreign policy making over a period of time. The question is, while it is true that all nations take care of their own national interests, are those interests consistent with the requirements of statehood under international law? A state is defined under the *Montevideo Convention on the Rights and Duties of States* (1933) as having a permanent population, defined territory, government, and capacity to enter into relations with other states. Using one or more of these elements of statehood, the author argues that there are deficiencies of the Philippine state in its approach to achieving and maintaining its national interests.

Under the definition of a state, having a defined territory is a requirement. In three instances, the Philippines has failed to enforce its rights as a sovereign nation. In dealing with other nations, the Philippines is supposed to preserve, protect and defend its territorial integrity. These occasions are: first, the debacle over our claim to Sabah. By all indications, documentary and historical accounts, that territory belongs to the Philippines. Second, China's encroachment in Philippine waters has resulted in the dismemberment of Philippine territory. Chinese actions in occupying Mischief Reef reduced the Philippine Exclusive

---

[107] Ridzwan Rahmat, *Philippines to deploy BrahMos missiles with new coastal defence unit*, **janes.com**, 17 January 2022.

Economic Zone (EEZ), one of its entitlements under UNCLOS. Worse was felt when China took control of Scarborough Shoal, a feature that is considered a dagger pointed at Zambales, if not at the metropolis, the National Capital Region.

The author chose the *Montevideo Convention* as a reference point for analysis for two reasons: first, it entered into force on December 26, 1934, before the Philippines became a commonwealth. The Philippines was therefore under the full control of the United States administration; second, the Philippines had no Department of Foreign Affairs at that time. Our foreign relations were handled by the U.S. Department of State. Since the U.S. is a signatory to the Treaty, the Philippines, by extension as an unincorporated U.S. territory, was bound by the agreement. Under the agreement, "...the state has the right to defend its integrity and independence, to provide for its conservation and prosperity."[108] It appears, therefore, that even before the arrival of UNCLOS on the international scene, or even the much earlier U.N. Charter, legal rights have already existed, accrued, and have been embedded in relations among nations. These are portions of customary law which have been recognized since time immemorial. In pre-World War II days, China was in disarray, occupied by the colonial powers, and was in no position to affect the territorial integrity of other states. As indicated by the *Montevideo Convention*, conservation of resources was already in the minds of the plenipotentiaries, thus foreseeing the limits on matters like the allowable catch of fish and other forms of food and valuable but finite items from the seas.

The Convention assures states, powerful and powerless, that they have the same rights and legal capacity in the

---

[108] *Montevideo Convention on the Rights and Duties of States*, Article 3, Montevideo. Entered into force on December 26, 1934. https://www.jus.uio.no/english/services/library/treaties/01/1-02/rights-duties-states.xml.

international arena. It says that, "The rights of each one do not depend upon the power which it possesses to assure its exercise, but upon the simple fact of its existence as a person under international law."[109]

A highly respected rule of international law is its abhorrence to the use of unnecessary or illegal force.[110] In China's occupation of Mischief Reef and Scarborough Shoal, there was no exchange of gunfire. But there was physical prevention of access by the Philippines to those features for as long as the Chinese forces were present there. There is no expectation by the Philippines, the international community, or any other country, that the Chinese will leave the occupied features in the foreseeable future.

Hypothesis

It is the hypothesis of the author that Philippine foreign policy should be more realist in its approach. By this the author means that the Philippines needs to be engaged in the singular quest for self-reliance in its relations with other countries. In foreign affairs, national security, and defense, a state should be able to stand on its own. Alliances or cooperative arrangements[111] are appurtenances to and support the national interest. But the innermost core of a nation's strength is its own inherent capability.

To support this hypothesis, the author intends to show the performance of the Philippines in the international arena. Hypothetical questions will arise: is the Philippines fulfilling its mandate to pursue the best interests of the Filipino people? Is it short of that aspiration? On the latter question, the opinion of the

---

[109] Article 4, *Ibid.*

[110] Article 11, *Ibid.*

[111] *Locsin meets India's External Affairs Minister*, **Manila Times**, February 21, 2022, https://www.manilatimes.net/2022/02/21/expats-diplomats/locsin-meets-indias-external-affairs-minister/1833618.

author is that the Philippines falls short in its compliance with the Constitution. How does it compare with other nations? Are its treaty commitments beneficial to the country? Is its national power consistent with the requirements of its laws and court decisions? Are those decisions enforceable, for the benefit of the country and the Filipino people?

Comparisons in the international community about the performance of their respective governments oftentimes involve that of the United States. But the United States is so advanced that its own challenges may not even be approximated by countries like the Philippines. It is so far ahead economically, politically and socially that the nature of its problems is beyond the reach, at least resource-wise, of third world countries. In the field of foreign relations, certainly, the realist school is alive and well in the United States. This is shown by the ideas and initiatives of statesmen like Henry Kissinger. In the post-war period, they had George Kennan, who envisioned a contained USSR. His policy prescription succeeded. The Soviet Union disintegrated. Yet, there are trends that the United States has to contend with: the push for global economic reform, a changing security landscape, America's energy revolution, and its valuable foreign policy asset as a cultural leader: its private sector.[112] It is in the international security universe that realism will pervade the atmosphere among all nations, including the United States.

Assumption

The study assumes that the Philippine system of government remains the same, that is, it is operating under the 1987 Constitution.

---

[112] https://weforum.org/agenda/2016/04/4-trends-that-will-define-future-of-us-foreign-policy/.

A good way of studying Philippine national interest and the game plan to attain state power is to begin with the Constitution. The Constitution acknowledges that the Filipino people are sovereign (Preamble). Article 1 defines the national territory. This article mixes conflicting concepts with the UN Convention on the Law of the Sea (UNCLOS). On the one hand, Article 1 says, "The waters around, between, and connecting the islands of the archipelago, regardless of their breadth and dimensions, form part of the internal waters of the Philippines." This provision of the Constitution runs contrary to the UNCLOS because of the general rule that the term "internal waters" refers to the landward side of the baseline of the territorial sea (UNCLOS, Article 8). Yet, we agree with UNCLOS more than with our own constitution. A case in which the Supreme Court ruled that the Philippine Baseline Law, Republic 9522, is constitutional is *Magallona versus Executive Secretary* (G.R. No. 187167, August 16, 2011). It is interesting to note the argument of the petitioners, who lost the case, that RA 9522, passed pursuant to UNCLOS, reduces Philippine maritime territory and opens the country's waters landward of the baselines to passage by all vessels and aircraft, undermining Philippine sovereignty and national security.

The constitutional imperative is the broad outline of the nation's legal, social, political, and economic framework. The foundation of any democratic government is its system of laws, foremost of which is its constitution. As the country's fundamental law, it is the basis of its national consensus and sets the extent and the limits of its laws, rules, and regulations. It keynotes the institutional make-up of the nation, including its economic moorings and social set-up. A constitution's length and level of detail is perhaps an indication of agreement on country-wide issues. If a constitution is lengthy, one sees that there is a requirement to commit specific provisions in writing, thus indicating the potential for contention if those provisions have not been written. On the other hand, if a constitution is short, and omits certain practices that are already deeply embedded in society (like the common-law system in some countries), there is greater likelihood that a state-wide agreement on details exists. This is so even if certain practices are not written down, thus showing a common belief in institutional arrangements that are taken for granted as given in society. One excellent example is the lack of a written constitution in some countries, like the United Kingdom.

The Constitution says that the government is composed of three branches: the executive, legislative, and judiciary. The Supreme Court is the highest tribunal in the land. The law is what the Supreme Court says it is, as it is the arbiter of disputes and its final word is not appealable.

Realist-Related Language in the Constitution

The Constitution, by definition, is state-centered. It has provisions that reflect the primacy of the state. For instance, it says that "The prime duty of the Government is to serve and protect the people. The Government may call upon the people to defend the State and, in fulfillment thereof, all citizens may be required, under conditions provided by law, to render personal, military or civil service."[113] Here we see a statement that makes the state the center of life of all citizens. Its protection and preservation, indeed its survival, is a duty of each and every member of that state. We may recall that survival of the state is a realist element of analysis.

As part of its state policies, "The State shall pursue an independent foreign policy. In its relations with other states, the paramount consideration shall be national sovereignty, territorial integrity, national interest, (underscoring supplied) and the right to self-determination."[114] Realists put national interest over and above other considerations in relation to their country. That same phrase

---

[113] Article II, Section 4, 1987 Constitution.
[114] Section 7, Ibid.

is used when the Philippines declares its policy with respect to nuclear weapons, thus, "The Philippines, consistent with the national interest, (underscoring supplied) adopts and pursues a policy of freedom from nuclear weapons in its territory."[115] On the economic front, the Philippines aspires to be on its own, as much as possible. Self-help is emphasized in that "The State shall develop a self-reliant and independent national economy effectively controlled by Filipinos."[116] The characteristics of being alone, without depending on anyone, is a realist element of freedom-loving and independent states.

Definition of Terms

The following terms shall be understood in the context described:

Bipolarity—the existence of two powers contending for supremacy in the international order.

Constitution–the fundamental law of the land.

EEZ—Exclusive Economic Zone "is an area beyond and adjacent to the territorial sea, subject to the specific legal regime established in Part V of UNCLOS, under which the rights and jurisdiction of the coastal State and the rights and freedoms of other States are governed by the relevant provisions of the Convention."[117]

Mutual Defense Treaty (MDT)--an agreement between the Philippines and the United States, signed in Washington, D. C., on August 30, 1951, whereby they pledged to each other that "an armed attack in the Pacific area on either of the Parties would be dangerous to its own peace and safety and declares that it would

---

[115] Section 8, Ibid.
[116] Section 19, Ibid.
[117] Article 55 **Specific legal regime of the exclusive economic zone**, Part V, UNCLOS.

act to meet the common dangers in accordance with its constitutional processes." (Article IV of the Treaty).

Multipolarity—the presence of many powers competing for prominence and influence in the international system.

National interest—the aspirations and goals of sovereign entities in the international arena.[118] As sovereign entities, states may include in this definition their possessions, tangible or intangible, such as territory, political autonomy, culture, and their populations.

NATO—North Atlantic Treaty Organization. A Western European security alliance that provides that an attack on one member country will be considered an attack on the others. Thus, those who are not involved in the armed conflict are obliged to assist by military means that beleaguered member.

Neorealism— a modification of the realist theory in that all countries belong to a system where each one has the same goal as the others, that is, to attain security under an environment of uncertainty and anarchy. Neorealism is a systemic theory where the distribution of power is the underlying determinant of the placement of states within the structure.

Realism—a theory of international relations that focuses on the state, on its own, and its attainment of national power pursuant to its interests.

SCS—the South China Sea. The body of water west of Palawan bordering China and Vietnam.

Security dilemma—a situation in which a country that strives to have a strong defense and national security apparatus appears to threaten another who, in turn, develops a powerful state, which

---

[118] Donald E. Nuechterlein, *National interest and foreign policy: a conceptual framework for analysis and decision-making,* published online on 26 October 2009 by Cambridge University Press.

compels the former to further strengthen itself, and the cycle goes on until, eventually, no one feels the security that they originally sought.

UNCLOS—the United Nations Convention on the Law of the Sea, 1982.

Unipolarity—"a system that features a highly asymmetric distribution of relative capabilities on a global scale, leaving one state significantly stronger than the rest and too strong to be counterbalanced."[119]

West Philippine Sea—WPS. The body of water is next to the baselines of the Philippines along Palawan, Zambales, and the Northern Provinces.

---

[119] Camilla T. N. Sorensen, **Is China Becoming More Aggressive? A Neoclassical Realist Analysis**, *Asian Perspective*, July-Sept. 2013, Vol. 37, No. 3 (July-Sept. 2013) p. 369, citing William C. Wohlforth, 1999, as source of definition.

Chapter 3: Methodology

Research Design

The work begins with a description of the realist theory. On the basis of the work of Morgenthau, a parallel study is made of the Montevideo Convention on the Rights of States and the Constitution of the Philippines and other laws and conventions, such as the UN Convention on the Law of the Sea. Here, an attempt is pursued whether the Philippines is realist enough. Has it sought power enough? Has it focused well in the preservation and pursuit of its national interest?

The overall design will be qualitative. It will be a case study of the Philippines in its policy execution in pursuit of its national interests. Sources of the study include policy statements, declarations,[120] issuances, speeches,[121] and other concrete pieces of evidence of the government approach to administration. News media is a source of empirical study. For example, if a ranking official states in a press conference that the Philippines will do everything it can to defend the Kalayaan Island Group (KIG) in the South China Sea, this statement can be validated in many ways: visibility in number and quality of troops and equipment stationed in the KIG, and the attention that they get from the national government.

Media items reflect social beliefs and behavior. As a source of information, they reveal attitudes towards the government and how the latter handles the national interest. Media can also give a hint how, in practical terms, realism dwells in day-to-day life.

---

[120] *Locsin: Arbitral Award PH's contribution to boosting legal order over the seas*, **ABS-CBN News**, Posted at Feb 23 2022 05:43 AM. This was a ministerial forum involving EU and ASEAN.

[121] Richa Noriega, *Philippines, Poland firm sign deal for 32 Black Hawk helicopters*, **GMA News**, Published February 22, 2022, 4:53 pm.

Another way of looking at the Philippine approach to achieve and protect its national interest is to survey one Supreme Court case that has something to do with foreign relations, such as dealing with national boundaries. In this case, it is the author's opinion that the law sought to be amended, RA 3046, gave a much wider breadth of territorial sea to the Philippines, 274,136 square nautical miles. On the other hand, the amending law, RA 9522, in compliance with UNCLOS, gave the country 32,106 square nautical miles. RA 9522 was sustained as constitutional by the Supreme Court.[122] As we know, in the territorial sea, the country exercises sovereignty and jurisdiction, while in the EEZ, the Philippines only has sovereign rights. The Treaty of Paris comes to mind. Is the decision consistent with policy, or the political aims of the government? Can the decision be relied upon in Philippine development objectives?

---

[122] *Magallona v. The Executive Secretary, G. R. No. 187167,* August 16, 2011. Figures are stated in the table that respondents submitted in their comment to the Supreme Court. This case is mentioned in the text of this study, *supra.*

Table 4: <u>Figures Submitted by Respondents</u>, cited in *Magallona v. The Executive Secretary*, described *supra*.

	Extent of maritime area using RA 3046, as amended, taking into account the Treaty of Paris' delimitation (in square nautical miles)	Extent of maritime area using RA 9522, taking into account UNCLOS III (in square nautical miles)
Internal or archipelagic waters	166,858	171,435
Territorial Sea	274,136	32,106
Exclusive Economic Zone		382,669
TOTAL	440,994	586,210

Let us focus on the figures on the territorial sea. Under RA 9522, the law disregarded the Treaty of Paris limits, in order to comply with UNCLOS. As a result, the Philippine territorial sea was reduced by a whopping 242,030 square nautical miles, an 88 percent reduction in area. Granting that we gained an enormous EEZ, a country only has sovereign rights over that body of water.

It is even subject to other rights by the international community, such as: freedom of navigation and overflight, and the right of transit passage. These rights of the international community are non-derogable and cannot be suspended. By contrast, in the territorial sea, a country has full sovereignty, subject only to the right of innocent passage of the vessels of other countries. And, the right of innocent passage can be suspended in cases of imminent threats to the national security of the coastal state. The author's view is that the Philippines lost too much in having a drastically reduced territorial sea.

As a theoretical study, the author will highly probably derive theoretical results. However, a sound policy is underpinned by a solid theoretical foundation. There is no question at all on the validity and reliability of works and statements because every input comes from a respected author or a governmental source. Whether it be a treaty, a speech or a court decision, there is no need to check the authenticity of such documents. News media are reflections of public opinion that make judgments on state performance.

The author assures the readers that there will be no ethics issues that will arise as a result of the study. If issues engender any debate at all, it will only be from an academic standpoint. It will not involve personalities. It will be purely issue-based.

Research Locale

The research focuses on the Philippine territory in relation to its location in Southeast Asia and the world. The South China Sea is one of the areas of contention between the Philippines and a few powers, especially China. The Philippines views the SCS as a place to prove its sovereignty and explore its resources for its growing population. There are two specific areas of concern for the Philippines, the return of both will restore the honor and dignity of the Republic. These are: Mischief Reef, a part of the

Philippines' Exclusive Economic Zone (EEZ) off the southern coast of Palawan, and Scarborough Shoal, also well within the EEZ of the country, west of Zambales.

According to a publication of the US government, "In the South China Sea, Beijing will continue to use growing numbers of air, naval, and maritime law enforcement platforms to intimidate rival claimants and signal that China has effective control over contested areas. China is similarly pressuring Japan over contested areas in the East China Sea."[123]

In connection with Philippine colonial experience, the capital cities of Washington, D. C. and Madrid can also be mentioned as part of the research locale. London may also be included because of the US-UK Agreement that delineated US jurisdiction over the Philippines.

Samples and Sampling Techniques

The samples to be used are generally academic work of scholarly individuals, the Philippine Constitution and contemporary events as reported in the media.

As mentioned, national interest is the gauge by which a country's activities are aligned with the wishes of its people as enshrined in its constitution. It is in the national interest of the Philippines to be self-sufficient, self-reliant and independent in all spheres of governmental activity.

There is nothing wrong with concessions from foreign countries. But if these concessions have been going on since we

---

[123] *Annual Threat Assessment of the U. S. Intelligence Community*, **Office of the Director of National Intelligence**, February 7, 2022, p. 6. This publication notes that "Information available as of 21 January 2022 was used in the preparation of this assessment." It is so indicated on the title page and page 2 of the assessment.

became an independent country, there must be something wrong. If these concessions are reciprocal or part of trade agreements, these are not inimical to the national interest. But if they are viewed as part of unilateral privileges bestowed on the Philippines, then our country would appear to not have matured over the years. It is still dependent on a foreign power. If it seeks special treatment from other countries, it becomes a beggar, and a potential pariah, in the international environment. "Donor fatigue" is often used as a term referring to nations whose assistance is always sought by dependent countries.

In one case, it was determined in a Senate hearing that the power grid of the Philippines is now under the control of China. This unbelievable occurrence arose out of the acquisition by China of 40% of the National Grid Corporation of the Philippines (NGCP) and reconfiguring the system of the corporation for its (China's) purposes.[124] Because of this development, the grid can be operated remotely by its system that is located in Nanjing, China.[125] The author believes that no self-respecting country should allow something like this to happen.

There are many pieces of evidence where the Philippines asserts its rights only tangentially, meaning not directly springing from its national power. These assertions are meant to put on record the Philippine position. The position is based on weakness, i. e., the statements reaffirm rights on the bases of international law, not on the strength of the nation. This is because the Philippines cannot issue ultimatums. It cannot give warnings. It can only make pleas before international audiences. Thus, we view diplomatic protests as totally useful.[126] The act of protesting is an

---

[124] EDITORIAL, *Power in the wrong hands*, **Philippine Daily Inquirer**/04:08 AM, November 26, 2019.

[125] *Ibid.*

[126] Paolo Romero, HEADLINES, *Diplomatic protests still effective vs China—experts*, **The Philippine Star**, December 7, 2021 | 12:00 am.

indirect way of showing our raw power, which we do not really have when confronting China. The diplomatic protests serve as part of the record so that we will not be placed in estoppel, meaning that we did not sleep on our rights.[127] In another instance, the media reported that gas drilling will reinforce our rights in the West Philippine Sea (WPS). Again, this is an indirect display of power, simply because we have nothing raw to display. It is as if, by drilling, the Chinese will be put on notice that they are violating our rights if they perform the same activity in the same place, something that they can do, and we cannot physically prevent them from doing so. According to Rep. Johnny Pimentel, "We are counting on Forum Energy Ltd.'s offshore drilling in Sampaguita to demonstrate that the Philippines is determined to enforce the undivided rights over Recto Bank's vast petroleum resources."[128] It looks incongruous that the government is depending on a private enterprise to assert a sovereign right that is within the ambit of the state's responsibilities under the Constitution.[129]

As a sovereign nation, the state is mandated to secure its borders, mark the metes and bounds of its territory and develop them for its people and their succeeding generations. In the case of the Philippines, the starting point is the Treaty of Paris of 1898. This Treaty ended the Spanish-American War that resulted in the transfer of jurisdiction over the Philippines, Cuba, and Puerto Rico to the United States. Puerto Rico was annexed by the United States, Cuba was released as an independent country, and the Philippines became a commonwealth that eventually led to its

---

[127] Gabriel Pabico Lalu, *Lacson asks DFA: What happened to 200 diplomatic protests vs China?*, **INQUIRER.net** / 03:38 PM December 06, 2021.Also reported by Hana Bordey, *DFA Philippines filed 241 diplomatic protests vs. China since 2016*, GMA News 7:11 pm, December 6, 2021.

[128] Neil Arwin Mercado, *Sampaguita gas drilling will 'forcefully assert' PH rights over WPS—solon*, **INQUIRER.net**/01:20 AM, December 07, 2021.

[129] 1987 Constitution, Article I, National Territory.

independence from the US in 1946. The Treaty of Paris map precedes all other drawings that show the land and water areas under Philippine jurisdiction. In the author's opinion, there is a policy failure on the part of the Filipino leaders early on not to have delimited Philippine territory so that it could have been laid out on a map that identified clearly what water and land areas are under the Filipino nation: its configuration, its dimensions, and the breadth and reach of its law. That omission was a misstep in policy making. When an individual buys or inherits real property, the first thing that he does is to delineate the boundaries with respect to his neighbors so that future territorial conflicts with them will be avoided. As it is true with individual persons, so it is also true among nations.

The first serious law that dealt with demarcation of national boundaries was in 1961, some 63 years after the effectivity of the Treaty of Paris of 1898. This law was known as RA 3046. Much later, in the 1990s, RA 9522 was passed to amend RA 3046. It is obvious that boundary determination for Philippine territory, even if it was provided for in the Constitution, did not enjoy priority among legislators for a long time. Toward the end of 2021, a draft law that intended to declare maritime zones hurdled the lower house.[130]

After the UNCLOS took effect on 16 November 1994, a long 96 years after the Treaty of Paris took effect, the Philippines, by law, sought to conform to the provisions of UNCLOS. It is not logical that a treaty (the Treaty of Paris), a part of the law of the land that took effect almost a century before a later treaty (UNCLOS), should conform to the latter.[131]

---

[130] Anna Felicia Bajo, *Bill declaring Philippines' maritime zones gets House's nod*, **GMA News**, December 6, 2021, 4:34 pm.

[131] For more detailed facts, see *Magallona v. Executive Secretary*, G. R. No. 187167, August 16, 2011.

Data Collection and Management Procedure

There might be no hard and fast data on this subject because geopolitics changes. But there must be objective criteria upon which to evaluate Philippine performance in its statehood since attaining independence. To enable the author and the readers to be on a common plane of understanding, let us focus on issues and developments on the West Philippine Sea. We can use it as a test case for leadership and management of the national interest where power, if any, is employed. The data is available in the media and open sources.

The West Philippine Sea: A Test of Leadership and Management

West Philippine Sea issues can be traced to the 1950s when a Filipino explorer and entrepreneur, nicknamed "Admiral" Tomas Cloma, traveled to the Spratly group of islands and claimed a portion of it on behalf of the Republic of the Philippines. He named them Freedomland. Much later, during his administration, President Marcos issued Presidential Decree No. 1596 in 1978, which constituted the Kalayaan Island Group (KIG). The KIG was made a municipality and placed under the administration of the Province of Palawan.

There was relative calm and stability in Palawan's governance because of national government support. In addition to political backing, the Armed Forces of the Philippines activated the Western Command, tasked to secure the features from encroachment by foreign forces. Around the late seventies, the Philippines had identified competing claimants to the islands and features occupied by it. In fact, the previously unoccupied atolls have begun to have foreign troops in them. These countries are: China, Vietnam, Malaysia, and Taiwan. Brunei is a claimant as well but it currently has not stationed troops anywhere in the Spratlys.

The KIG has a sense of normalcy in its status as a regular town of Palawan. It has an elected local government led by its mayor and municipal council. It has a school. It has a refurbished airstrip and a small port that can receive vessels, civilian and military. The runway is capable of hosting cargo aircraft from the Philippine Air Force.

In 1995, China occupied Mischief Reef, one of the features under the control of the Philippines, and located within its Exclusive Economic Zone (EEZ). This Chinese action shocked the Philippines. China countered that the occupation of the feature was made to establish a shelter for fishermen in distress. This was belied in the following years when sophisticated antennas and heavy construction indicated that Mischief Reef had the new character of a military outpost. The Philippines tried to rally the international community behind it but with little effect. Chinese occupation and militarization of Mischief Reef continues to this day.

In 2012, China blocked access of a Philippine vessel to Scarborough Shoal, a feature about one hundred miles west of Zambales Province of the Philippines. It is well within the EEZ of the Philippines and has been used as a target area of aircraft from the U. S. and the Philippines when American troops were stationed in Philippine bases. A tense standoff ensued, which eventually led to the withdrawal of the Philippine Navy vessel from the area. The Chinese have stayed in the feature since then.

Every ASEAN Summit since the Chinese occupation of Philippine-held features has resulted in a statement of concern on freedom of navigation. ASEAN also showed its distress on the militarization of the features and their conversion into artificial islands for military purposes. But for the first time in its history, ASEAN did not come up with a closing communique in 2012. The chair of the group at that time was Cambodia. It is believed that the chair succumbed to Chinese pressure not to issue a statement at the end of the meeting. The tradition was broken. A fissure within ASEAN has been observed by the international community. Despite the continuing negotiations within ASEAN for a Code of Conduct for parties in the South China Sea, there is little optimism that the Code will soon emerge, or if it does, it will be observed. Power is still the arbiter of issues in the South China Sea. Thus, a realist approach is called for.

It was against this backdrop that the Philippines decided to bring the matter to international arbitration. On January 22, 2013, the Philippines instituted arbitral proceedings against the People's Republic of China under Annex VII of the UNCLOS.[132]

On October 29, 2015, the Tribunal rendered an Award on the basis of Jurisdiction and Admissibility.[133] The following declarations were made in the award:

- The Tribunal held that both the Philippines and China are parties to the Convention and bound by its provisions on the settlement of disputes.

- The Tribunal also held that China's decision not to participate in the proceedings does not deprive the Tribunal of jurisdiction and that the Philippine decision to commence arbitration unilaterally was not an abuse of the Convention's dispute settlement procedures.

---

[132] *Permanent Court of Arbitration (PCA) Press Release,* **"Arbitration between the Republic of the Philippines and the People's Republic of China: Arbitral Tribunal Establishes Rules of Procedure and Initial Timetable,"** August 27, 2013. https://pcacases.com/web/sendAttach/227. Annex VII refers to the provisions on arbitration annexed to the UNCLOS that parties may refer to for procedural requirements, jurisdiction, and grounds for institution of proceedings.

[133] *PCA Press Release,* **"The Tribunal Renders Award on Jurisdiction and Admissibility; Will Hold Further Hearings,"** October 29, 2015. https://pcacases.com/web/sendAttach/1503.

- Upon review of the claims submitted by the Philippines, the Tribunal rejected the argument in China's Position Paper that the dispute is actually about sovereignty over the features in the South China Sea and therefore beyond the Tribunal's jurisdiction.

- The Tribunal also rejected the argument by China in its Position Paper that the Parties' dispute is actually about the delimitation of a maritime boundary between them and therefore excluded from the Tribunal's jurisdiction through a declaration made by China in 2006. On the contrary, the Tribunal held that each of the Philippines' Submissions reflect disputes between the two States concerning the interpretation or application of the Convention.

- The Tribunal held that no other States are indispensable to the proceedings.

On July 12, 2016, the Tribunal rendered its Award,[134] as follows:

The Award is classified into three major headings: maritime entitlements, status of South China Sea (SCS) features, and the legality of certain actions of China in the SCS.

On maritime entitlements, the Arbitral Award declared that the UNCLOS superseded any historic rights, or other sovereign rights or jurisdiction, in excess of the limits imposed by UNCLOS. Thus, China's claims to historic rights, sovereign rights or jurisdiction in areas of the "nine-dash line" are contrary to UNCLOS and without lawful effect in so far as they exceed the geographic and substantive limits of China's entitlements under the Convention. This means that China cannot claim entitlements such as exclusive use of marine resources in areas of the "nine-dash line" that exceed geographic and substantive limits, for example,

---

[134] *Award on the South China Sea Arbitration*, July 12, 2016. https://pcacases.com/web/sendAttach/2086.

the Reed Bank area. The Award also declared that Mischief Reef and the Second Thomas Shoal are within the Exclusive Economic Zone (EEZ) and Continental Shelf (CS) of the Philippines.

On the status of SCS features, the Arbitral Award declared that the following features as high-tide elevations:

- Scarborough Shoal, also known as Bajo de Masinloc;

- Gaven Reef North;

- McKennan Reef (Chigua Reef);

- Johnson Reef (Mabini Reef);

- Cuarteron Reef (Calderon Reef; and

- Fiery Cross Reef (Kagitingan Reef)

These features do not generate any entitlement to a 200-nautical mile EEZ and CS. These features generate a 12-nautical mile territorial sea only. Therefore, even if China occupies any of these high-tide elevations and has developed these as artificial islands, China still cannot claim any EEZ or CS entitlements for these features. China occupies all the six features listed above. Features numbered 1, 3 and 4 are within the EEZ and CS of the Philippines.

The Arbitral Award declared the following features as low-tide elevations:

A. Subi Reef or Zamora Reef;

B. Gaven Reef South;

C. Hughes Reef;

D. Mischief Reef or Panganiban Reef; and

E. Second Thomas Shoal or Ayungin Shoal

These features do not generate entitlement to a 200-nautical mile EEZ and CS and are not subject to appropriation. Thus, even if China occupies any of these low-tide elevations, and has built artificial islands on them, China cannot claim any territorial sea, EEZ or CS entitlements for these features. More importantly, China cannot appropriate any of these low-tide elevations even if it occupies them. Features A, B, and D are occupied by China. Under UNCLOS, low-tide elevations are considered part of the CS of the coastal state. Features D and E are inside the EEZ and part of the CS of the Philippines.

In addition, the Arbitral Award emphasized that the high-tide elevations in the Spratly Islands are not fully-entitled islands, and as such, these high-tide features do not generate EEZ or CS entitlements. Therefore, no EEZ or CS entitlement is generated by any feature claimed by China that overlaps with Philippine entitlements in the areas of Mischief Reef, Second Thomas Shoal, and Reed Bank.

On the legality of certain actions by China in the SCS, the Arbitral Award declared that China breached its obligations under UNCLOS and related international conventions, thus:

- China's operation of its official vessels at Scarborough Shoal unlawfully prevented Filipino fishermen from engaging in traditional fishing in the area.

- China's operation of its law enforcement vessels on April 28, 2012 and May 26, 2012 in the Scarborough Shoal area, which created serious risk of collision with Philippine ships and personnel, violated several provisions of the Convention on International Regulations for Preventing Collisions at Sea (COLREGS).

- China's operation of its maritime surveillance vessels to tolerate and protect Chinese-flagged vessels fishing in the Philippine EEZ at Mischief Reef and Second Thomas Shoal.

- China's construction of artificial islands, installations and structures at Mischief Reef or Panganiban Reef without Philippine authorization violates UNCLOS provisions on the EEZ and CS entitlements of the Philippines because the Reef in question is a low-tide elevation located in the Philippine EEZ and CS.

- China's land reclamation and construction of artificial islands, installations and structures caused severe and irreparable harm to the coral reef system at: (1) Mischief Reef or Panganiban Reef, (2) Johnson Reef or Mabini Reef, (3) Cuarteron Reef or Calderon Reef, (4) Fiery Cross Reef or Kagitingan Reef, (5) Subi Reef or Zamora Reef, (6) Gaven Reef North, and (7) Hughes Reef.

- China's tolerance of and failure to prevent harvesting of endangered species and giant clams in a manner that was severely destructive of the coral reef ecosystem in the SCS.

The Tribunal has recognized artisanal fishing. It said that the fishing activities that have been traditionally performed by neighbors in the area are not prohibited. Thus, the fishing rights of nationals within the region are permitted to continue in recognition of many years of customary practice.

From the decision of the Arbitral Tribunal, it would seem that the matter has been settled, but that is not the case. From the very beginning, the Chinese side refused to recognize the jurisdiction of the Tribunal. In fact, China refused to participate in the proceedings. The perception is that the Philippines and China are in a stalemate. The real situation is that China refuses to comply with the ruling of the Arbitral Tribunal. With its presence on and its militarization of the features, the Philippines cannot enforce the favorable judgment of the Tribunal.

The issues involved in the West Philippine Sea, like sovereignty, territorial jurisdiction, and national security will test the leadership of any head of state. It will also be an opportunity for Philippine leaders to have their management skills honed, or exposed, for lack of it. It is a challenge fit for the very top captains of the ship of state.

With the election of new leaders in the Philippines, the architecture or the narrative, if it has to be changed, of the discussions on the West Philippine Sea will be determined by the talents and acumen of the new set of policy makers. The author recommends that the new leaders should be realists: that they have the national interest of the Philippines first and foremost in their minds.

At least six experts on international relations are not convinced that potential leaders of the Philippines have sufficient expertise in foreign policy to lead the nation.[135] This was shown by the performance of the candidates in interviews in connection with the coming presidential elections.

---

[135] Alyssa Nicole O. Tan, Reporter, and Jasper Emerald G. Tan, *Analysts decry unclear foreign policy direction*, **BusinessWorld**, March 6, 2022, 7:40 pm.

Chapter 4: Results and Discussions

Theories of International Relations in the Philippine Context

Theories	Philippine Context
Idealism	The goal of every country in the world. It is not realizable because of its lofty nature.
Liberalism	It comes and goes, depending on state leadership.
Rational Choice	It is assumed in any type of leadership.
Constructivism	Requires sophistication by the population due to varying interpretations of relationships, events and perceptions.
Neorealism	Closest to scientific approach. Systemic view still unable to explain certain international phenomena, e. g., fall of the Soviet Union.
Realism	Anarchy and national interest are given in the international arena. Power projects the national interest. The Philippines should have a realist foreign policy.

Following is a consideration of the *Pros* and *cons* of the theories of international relations that have been previously discussed:

Idealism. *Pro:* idealism is essentially everything that is good and beneficial. It is utopian. It includes things like contemplative activities that Aristotle has expressed advocacy for.[136]

Con: it is simply impossible to achieve as there are imperfections in life. No such thing as ideal in international relations. In fact, the reality of human relations, including its frailties, is one of the reasons that realism is so attractive as a theory.

Liberalism. *Pro:* considered by the United States to be "the golden age of international politics—a period of relative international stability, a period when systemic or great-power wars have been conspicuous by their relative absence, a period of enormous international prosperity, and a period in which individual rights and democratic values have achieved unprecedented gains."[137] Brands added that, "insofar as the United States benefits from a world that is more peaceful, more stable, more economically open and prosperous, and more respectful of individual rights and democratic values, it would seem that the liberal order has been a good thing for America."[138]

Con: there are those who believe that the liberal order has not benefited the United States and the countries that subscribe to it, such as those in Western Europe. They say that "the very

---

[136] Gavin Lawrence, *Aristotle and the Ideal Life*, **The Philosophical Review,** Jan., 1993, Vol. 102, No. 1 (Jan., 1993), p. 34. Published by Duke University Press on behalf of Philosophical Review. Stable URL: https://www.jstor.org/stable/2185651.

[137] Hal Brands, *American Grand Strategy and the Liberal Order: Continuity, Change, and Options for the Future*, **RAND Corporation** (2016) p. 27. http://www.jstor.com/stable/resrep02400.

[138] *Ibid.* p. 28.

contours of the liberal order actively damage American well-being and that promoting such a system merely allows others—from illegal immigrants to selfish industrial democracies—to profit at America's expense."[139]

Rational Choice. *Pro*: The rational actor proceeds from a "cool and clear-headed ends-means calculation after considering all possible courses of action and carefully weighing the pros and cons of each of them."[140] Another favorable factor defines, "a rational actor is one who, when confronted with 'two alternatives which give rise to outcomes...will choose the one which yields the more preferred outcome.'"[141]

Cons: On the particular subject of deterrence, as an example, "classic criticisms of deterrence theory turn on the charge that governments simply lack the necessary rationality to make it work, that they are particularly subject to irrationality (underscoring supplied) in times of intense crisis or actual attack."[142] There are also analysts who say that rationality may not be explainable by logical inference. For instance, in understanding Hitler's policies and behavior, one does it "Simply by understanding his goals. In other words, preferences are given and either actual or optimal behavior deduced. The question of what preferences and/or perceptions an actor *should* have is considered not particularly relevant for developing explanatory or predictive

---

[139] *Ibid*. p. 27.

[140] Frank C. Zagare, *Rationality and Deterrence*, **World Politics**, Jan., 1990, Vol. 42, No.2 (Jan., 1990), Cambridge University Press, p. 239. https://www.jstor.org/stable/2010465 citing Sidney Verba in his work entitled, "Assumptions of Rationality and Non-rationality in Models of the International System" in Klaus Knorr and Sidney Verba, eds., *The International System: Theoretical Essays* (Princeton, NJ: Princeton University Press, 1961), p. 95.

[141] *Ibid*. p.240, citing Duncan R. Luce and Howard Raiffa, *Games and Decisions: Introduction and Critical Survey* (New York: Wiley, 1957), p. 50.

[142] *Ibid*. p. 239, quoting Patrick Morgan.

theories of behavior."[143] Because of the belief that a rational choice is arrived at with deliberation and weighing of options, it has to be accepted that a rational choice "does not connote superhuman calculating ability, omniscience, or an Olympian view of the world."[144]

Constructivism. *Pro*: "Constructivists have convincingly shown the empirical value of their approach, providing new and meaningful interpretations on a range of issues of central concern to students of world politics."[145]

Con: Constructivism, surely, has its defects. One of them, and perhaps the most important of all, is that "constructivist theorizing is in a state of disarray."[146] At best, constructivism can only be considered an approach to research and study. It is still far from becoming a theory of its own. On the whole, constructivism can show promise as a way of looking at things differently from each side of the analysis. For instance, the manner the West looks at Russia drastically varies in the way it looks at, say, France. The former is not a friend or an ally. The latter is a pillar of Western civilization, so to speak. This perception is true even if both are not doing anything threatening to each other. The state of mind is a valuable tool of analysis. While it seems clear that constructivism, to date, requires an in-depth program of research, it has its broad

---

[143] *Ibid.* p. 242.

[144] Ibid. p. 243, citing Glenn H. Snyder and Paul Diesing, *Conflict Among Nations: Bargaining, Decision Making and System Structure in International Crises* (Princeton, NJ: Princeton University Press, 1977), chap. 5.

[145] Jeffrey T. Checkel, reviewing *The Constructivist Turn in International Relations Theory*, Reviewed Works: *National Interests in International Society* by Martha Finnemore; *The Culture of National Security: Norms and Identity in World Politics* by Peter Katzenstein; *Norms in International Relations: The Struggle Against Apartheid* by Audie Klotz, **World Politics**, Jan. 1998, Vol. 50, No. 2 (Jan. 1998), pp. 324-348, Cambridge University Press https://www.jstor.org/stable/25054040.

[146] *Ibid.*

promise, as it advances learning in "its own puzzles that concentrate on issues of identity in world politics and the theorization of domestic politics and culture in international relations theory."[147]

Neorealism. Neorealism is also known as "structural realism." As its name implies, it is closest to classical realism, except that in neorealism, there is a "system" in which the nations of the world operate. The system moves along levels: the unit (or state) level and the systemic level. Neorealism is also concerned with the distribution of power in the international system: unipolarity, bipolarity or multipolarity. In classical realism, the state by itself is the unit of analysis. Other than this basic difference, the analytical tools are practically the same: anarchy in the international system, self-help among nations, primacy of national interest, polarity in the power structure, and so on.

Pro: Neorealism focuses on the independence of states, especially in the national security arena. As Kenneth N. Waltz said: "If I depend more on you than you depend on me, you have more ways of influencing me and affecting my fate than I have of affecting yours."[148] To each his own is the guiding light of states. Sovereignty is an indispensable attribute of statehood. The concentration on "distribution of national capabilities" is keenly watched as well as the "balancing of power" between and among them, which repeats itself in the international scene.[149] The theory accurately predicts that once disturbed, the balance of power will someday be reinstated.[150]

---

[147] Ted Hopf (fn. no. 11), p. 172.

[148] Kenneth N. Waltz, *Structural Realism after the Cold War,* **International Security,** Summer, 2000, Vol. 25, No. 1 (Summer, 2000), The MIT Press, pp. 15-16. https://www.jstor.org/stable/2626772.

[149] *Ibid.* p. 27.

[150] *Ibid.*

Con: While the theory may be able to state what will happen in, say, balance of power situations, it cannot give the timing as to when this will occur. This is a major shortcoming of the theory.[151] Some analysts say that the realist theory did not succeed in predicting the fall of the Soviet Union. This alleged defect is an indication of the incomplete character of the theory. The defenders of the theory assert that it "deals with the pressures of structure on states and not with how states will respond to the pressures."[152]

On balance, Waltz has this to say to those who declare that the neorealist theory no longer works because the conditions under which it is expected to operate have changed: "Changes *of* the system would do it; changes *in* the system would not."[153] In effect, Waltz is saying that the international community still has the same system. The changes that we have seen are made only within that same system.

Realism. Also referred to as classical realism, this is the school of thought where the analysis of international events, situations, developments, and relationships among nations puts emphasis on national interest and national power.[154]

Pro: Morgenthau was the pioneering author on this subject in its modern incarnation. It is known that even the ancient Chinese states had some form of realist doctrines in their political philosophy. So did the Greeks and the Romans. What Morgenthau established was the basics of the analytical framework, in which he

---

[151] *Ibid.*

[152] *Ibid.*

[153] *Ibid.* p. 6.

[154] Robert Jervis, *Hans Morgenthau, Realism, and the Scientific Study of International Politics*, **Social Research**, WINTER 1994, Vol. 61, No. 4, Sixtieth Anniversary 1934-1994: The Legacy of Our Past (WINTER 1994), The Johns Hopkins University Press, p. 855. https://www.jstor.org/stable/40971063.

traced "the necessary evil in politics to human nature"[155] exemplified by "the desire for power or *animus dominandi.*"[156] Although Morgenthau believes that morality is generally separate from national interest and power, "a degree of moral consensus among nations is a prerequisite for a well-functioning international order."[157]

Con: It can be said that realism was the forerunner of neorealism. The latter is more scientific as it borrowed concepts from the economic theory of the firm (or the market).

Looking at all angles, realism is familiar to all political scientists and international relations practitioners. For this reason, it is good to start from it. It would therefore benefit Filipino policy makers, academics and diplomats to master its basic principles because it is from realism that one can branch out into the sub disciplines of international relations theories.

Significance of the Theories for the Philippines

The six theories enumerated here can play roles in analyzing Philippine foreign policy. It is the position of the author that, to develop a deep understanding of the Philippines' position in the international arena, it has to have a keen knowledge of what it wants: its interests as a country. At the same time, the Philippines should have the power to attain and maintain its interests. Here, a real and honest-to-goodness inventory of its physical and non-physical assets needs to be made in order to assess the country's resources, strengths, and weaknesses. On top of these, the Philippines has to generate national consensus to arrive at policy

---

[155] *Ibid.*, p. 868.

[156] *Ibid.* Jervis cites Morgenthau himself as the source in Morgenthau, 1946, p. 192.

[157] *Ibid.* p. 869.

decisions and approaches with minimal or no dissension. It is here that the realist approach comes to the picture.

Philippine and Chinese Actions re: the Arbitration Case on the SCS

The situation in the South China Sea is a crucial example of how the Philippines should protect its national interest. The author suggests that the national leadership is short of the realist standards that he advocates. In contrast to Chinese President Xi Jinping's[158] claim of taking the moral high ground, President Duterte appears to be conciliatory most of the time. President Duterte's actions show clear instances of conflict avoidance, if not obeisance. In one case, the Department of Energy (DOE) ordered a halt to the ongoing search for oil in the West Philippine Sea after the President publicly stated that "the country had no choice but to comply with a supposed joint exploration deal with China to avoid any conflict."[159] Instead of asserting Philippine rights and encouraging foreign investments, the move was a step backward and not conducive to a climate favorable to foreign investors.

On the same issue of oil exploration, the DOE reportedly "asked the Cabinet security, justice and peace coordinating cluster to immediately allow oil exploration activities in the West Philippine Sea."[160] This action is an indication that the executive branch has no clear direction on the matter. It is generally known, not only by Filipinos, but also by other Southeast Asians, if not the entire world, that the supposed oil exploration activities are within the Philippines' EEZ. Indeed, some of them are even conducted in

---

[158] *Xi Jinping at Boao 2022: Cooperation "biggest strength" in achieving bright future for humanity,* **CGTN,** 21 April 2022.

[159] Kyle Aristophere T. Atienza, Reporter, *Halt on WPS oil search seen to discourage investments,* **Business World Online,** April 21, 2022, 12:33 am.

[160] Helen Flores, *DOE seeks Cabinet cluster OK for West Philippine Sea oil exploration,* **Philstar Global,** April 20, 2022, 12:00 am.

the territorial sea. As an important component of the executive branch, the DOE should have deeply analyzed the situation and created SOPs and immediate recommendations whenever serious issues that need action arise. The SCS dispute has been ongoing for some time, long before the 1995 occupation of Mischief Reef by China. By this time, the Philippines should have adopted clear guidelines on how to react to breaking situations, without compromising its national interest. What is obvious is that the Philippines cannot do anything against China. The reason is that it has not developed its power to the point that it can inflict upon China serious damage in case an actual conflict arises. It depended on the United States for its external defense needs. Despite this need for US forces to defend the country, the Philippines ejected the American bases in 1991. Having secured victory in the arbitration case in the Hague on July 12, 2016, the Philippines did not capitalize on it and China never recognized the decision. Thus, the Philippines was not a realist: it did not put its national interest on top of its priorities. Meanwhile, there is denial and pretense that the relationship with China is normal. The Chinese side is well aware of the Philippine reluctance to push the limits of the relationship. "Chinese Ambassador Huang Xilian said that the recent telesummit between President Duterte and Chinese President Xi Jinping reflected the strong friendship between the two leaders. Both leaders affirmed that the improving bilateral relations should continue even beyond the outgoing President's term."[161] The normalcy implied in the telesummit exchange is not close to the truth.

It is a sad commentary to talk about Philippine performance in armed or unarmed conflicts within its territory

---

[161] Tristan Nodalo, *EXCLUSIVE: Chinese envoy hopes next PH leader will continue "friendly policy," keep momentum of PH-China ties.* **CNN Philippines**, April 12, 2022, 9:52:21 pm.

since the revolution against Spain. Apart from the glorious declaration of independence by President Aguinaldo on June 12, 1898, defeat after defeat ensued. In the Second World War, the Philippines had debacles in Bataan, Corregidor, and other parts of the country. This later led to the eventual fall of the entire Philippines to Japan. After independence, in the 1960s, the country lost Sabah. Much later, it lost Mischief Reef, followed by Scarborough Shoal. It seems that the Philippines was not ready to be a full-fledged nation because of its inability to defend its territory.

One wonders what the Chinese think of the Philippines and of Filipinos. An example of an incongruous situation was when a Chinese vessel entered Philippine territorial jurisdiction. The media reported that "After a recent intrusion of a Chinese navy ship in Philippine waters, President Rodrigo Duterte and Chinese President Xi Jinping pledged to exert all efforts to maintain peace, security, and stability by exercising restraint, dissipating tensions, and working on a mutually agreeable framework for functional cooperation."[162] It was the Chinese vessel that intruded into Philippine waters, yet, both countries are calling for restraint and cooperation! The Philippines should not be lulled into so-called cooperative arrangements while unable to recover the lost features belonging to it. While China continues to enjoy the features, like having them fortified for their purposes, and using them as permanent air stations, "Duterte has repeatedly defended himself from criticism by insisting that the Philippines cannot reclaim Chinese-occupied areas in the West Philippine Sea without violence and bloodshed."[163] Faint heart and military

---

[162] Vito Barcelo, *Duterte, Xi call for restraint*, **manilastandard.net**, April 10, 2022, 2:00 am. The "telesummit" between President Duterte and President Xi where the pledge was made was covered by Mauro Gia Samonte, *What is the Xi-Duterte deterrence for war?*, **The Manila Times**, April 10, 2022.
[163] MJ Blancaflor, *Duterte, Xi commit to boost PH-China ties amid WPS dispute*, **Daily Tribune**, @tribunephl_MJB, April 9, 2022, 12:29 pm.

weakness of the Philippines has emboldened China first, in not recognizing the results of the 2016 Arbitration Case, and, its continuing flouting of Philippine jurisdiction, as in areas within the Philippine EEZ, such as the Recto Bank.[164] This area, like much of the South China Sea, reportedly has huge petroleum deposits.

The Westphalian System, Its Adoption and Implications for China and the Western World

Of all the members of the United Nations Security Council, only China has no European tradition in diplomacy. It had its own. But with the arrival of Europeans as the sixteenth century ended, China gradually evolved into a nation-state along the European mold, at least in its diplomatic practice. It had no choice. It had to adapt to modern ways.

The Treaty of Westphalia was a result of decades of war in Europe. It resulted in the recognition of boundaries, statehood of countries, and jurisdictions of independent territories. Eventually, passports and visas were to be required in entering and exiting nation-states. Towards the mid-1600s, there appeared on the scene of international affairs an updated, uniform manner of receiving envoys, conducting truce among warring parties, and diplomatically resolving conflicts. The changes that occurred affected Europe. It is that system that was in effect when European powers forayed into Chinese territory later. After its imperial collapse, China embraced Western diplomatic ways, abandoning the hierarchical character of its officialdom. The major implication of it is that more than a century later, China would rise and aspire to beat the West in its own game.

The significant events that led Europeans to adopt the Westphalian system were the incessant wars and religious conflicts

---

[164] Jarius Bondoc, *Kahinaan ng loob, isip sinasamantala ng China*, **Pilipino Star Ngayon**, March 29, 2022, 12:00 am.

that governed relations among European empires. The recognition of states became essential and the right of self-determination of territories transformed into a demandable right against the major powers of the time. Beyond Westphalia, China was a most logical starting point. For its part, China moved toward Westphalian principles as its peripheries began to fade away from its authority. The hierarchy and the tributes were gone.

Ironically, after the end of the Cold war, China has become "a staunch defender of the Westphalian order."[165] The Westphalian principles of non-interference in the internal affairs of other states greatly benefited China as it began to prosper politically and economically. It is evident, however, that "as China grows in geopolitical importance, its own strategic interests relating to interventionism will presumably also evolve."[166] The South China Sea picture is a case in point.

So far, the Westphalian system has worked in the international community. Nation-states have all complied with its rules. "Specifically, global relations among states and corporations increasingly rely upon coordination and collaboration rather than commands."[167] Orders from above, within a hierarchical and tributary state, which China was, cannot flourish. The fall of its hierarchical (the Middle Kingdom syndrome) and tributary system forced China to join the Westphalian Order.

---

[165] Yongjin Zhang, Footnote No. 33, p. 63.

[166] G. John Ikenberry and Amitai Etzioni, *Point of Order: Is China More Westphalian Than the West?* **Foreign Affairs**, November/December 2011, Vol. 90, No. 6 (November/December 2011), Council on Foreign Relations, p. 176. https://www.jstor.org/stable/23039640.

[167] Kurt Burch, *Changing the Rules: Reconceiving Change in the Westphalian System,* **International Studies Review**, Summer, 2000, Vol. 2, No. 2, Continuity and Change in the Westphalian Order (Summer, 2000), p. 181. Wiley on behalf of The International Studies Association https://www.jstor.org/stable/3186432.

There are certain takeaways that are relevant to the Westphalian system that are in effect. These are basic requirements in understanding the international system. Burch, quoting Young, says that "the Westphalian system of interacting sovereign states comprises a 'governance system,' 'an institution that specializes in making collective choices on matters of common concern to the members of a distinct group.'"[168] The Westphalian system is an institution, and it "exhibits 'persistent and connected sets of rules, formal and informal, that prescribe behavioral rules, constrain activity, and shape expectations.'"[169]

Indeed, the two major Vienna Conventions governing relations between states, have been offshoots of the principles enshrined in the Westphalian Order.[170] It can be said that the Westphalian system directly resulted in the uniform way of conducting diplomacy. Apart from the occasional violations of sovereignty among and between members of the international community, there appears to be no logical system that can replace it in the foreseeable future.

---

[168] *Ibid.* p. 185, citing Young, Oran. (1994) **International Governance: Protecting the Environment in a Stateless Society**. Ithaca, NY: Cornell University Press. p. 26.

[169] *Ibid.* quoting Keohane, Robert O. (1990) *Multilateralism: An Agenda for Research*. **International Journal** p. 732.

[170] The Vienna Convention on Diplomatic Relations, 18 April 1961, and the Vienna Convention on Consular Relations, 24 April 1963, Austria.

Chapter 5: Summary, Conclusions and Recommendations

The author argues that some of the schools of thought, or philosophical theories in international relations, namely: Idealism, Liberalism, Rational Choice, Constructivism, Neorealism and Realism have their own strengths. Realism, however, is arguably the most basic of all analytical tools. It is humbly proposed that Philippine foreign policy and international relations be viewed from the point of view of realism. This suggestion arises from the following summary of the findings of this study.

Summary of Findings

1. The Philippines has not been able to advance and protect its national interests in the international arena. This can be seen in the lackadaisical way in which it has treated its national boundaries. The Treaty of Paris gave to the Philippines a wide expanse of internal waters and it did not mark them or show signs of exercise of sovereignty. The Treaty of Paris was in force in Philippine waters more than fifty years before any of the talks that led to the UNCLOS. The Philippines also did not take care of its Sabah territory. It was lost by default, inaction, and lack of resolve.

2. The Philippines has not been able to develop its national power in order that it will attain the capability to protect its national interests. Take the case of the South China Sea issue. To date, the Philippines has been consistently avoiding confrontation with China. The usual reason given is that the Philippines does not want war. The other is that the Philippines is a peace-loving nation. But one of the requirements of statehood is territory, and having the capability to hold, defend, and protect that territory. The way the country is behaving with respect to the features that are within its EEZ and territorial waters, it would not be surprising if it is again the subject of dismemberment in the future. There are forthcoming generations of Filipinos. They need food, shelter, clothing and the necessities of life. Should the national leadership content itself with having reduced

fish production in the Scarborough Shoal because it cannot take on China (the way Vietnam does in other parts of the SCS), then the future young Filipinos will not be proud of their forebears. The reasons is that the leadership of the Philippines now and in the recent past is bequeathing unto them fewer resources, like fish, than the older generations had available to them. As a major source of protein, fish are greatly needed as the Filipino population grows in the coming years.

3. In its foreign affairs, it is the author's view that realism as an international relations theory be used to approach its dealings with other nations. Under realism, there is no central authority among countries in the world. There are only two major elements that matter: national interest and power. These two imperatives should be the guiding lights of the country's leadership.

4. The Westphalian system is here to stay. In Europe, after countless religious and monarchic wars in the seventeenth century, borders and nations have emerged from the ashes. This development gave rise to independence and self-determination of states. Despite occasional calls for world government, there is still no substitute for the Westphalian system. Given this situation among countries, individualism and security are the hallmarks of the game. Power and self-interest have to be achieved to secure the nation. If a country does not protect itself, no other country will assume that obligation. Since all countries have feelings of insecurity along their borders, it is incumbent upon each one of them to build, develop, and enhance the power necessary to achieve their own national security.

5. Looking back at Philippine historical experience in the management of its boundaries, it appears that it was deficient in its preparations for statehood after independence. The author makes this finding in view of the loss of Sabah, Mischief Reef and Scarborough Shoal. The remedies to these defects will be suggested in the recommendations of this study.

Conclusions

The realist school of thought in foreign relations points to power as the determining factor in a state's stature in the international system.

The Philippines is a middle-level country at this point in its history. It has to project itself above and beyond its current economic development so that it will find its appropriate place in the international community.

By doing this study, the author aspires to assist his country in putting it on the right track. The author's view is that the Philippines was not ready for real and actual independence even after several decades of attaining it on July 4, 1946. The reason is that it is unable to seriously defend its territorial integrity, one of the requirements of statehood. The country is not in full control of its territory. Sabah, Mischief Reef and Scarborough Shoal are stark reminders that the Philippines' sovereignty is not complete. Foreign powers have demonstrated that they can, and indeed have, interfered with its territorial boundaries.

The author enjoins all Filipinos to look at the neighbors of the Philippines to make the case that the country is probably not in its optimum capacity to achieve its desired national power. Even with its alliances and treaty commitments from friendly states, the Philippines appears short of its ideal national posture in the region. In the international arena, gone were the days when the Philippines held the Presidency of the General Assembly of the United Nations. It had commanded the respect of its neighbors and had commendable economic growth. As a student of realist theory, the author hopes that the Philippines will view the anarchical nature of the international system as an opportunity, not a liability, in its search for three goals: a realist-oriented foreign policy, a formidable defense posture, and the support base of these two goals, a robust and prosperous economy. Then, and only then, could the Philippines regain its golden years in the global arena from where it gained fame and admiration by the world community. From Korea to Congo and the Golan Heights, the Filipino distinguished himself. It is time for the Filipino leadership to cement that role, not only for future generations to cherish but also for the world community to emulate.

Recommendations

As the Filipino people have witnessed the weaknesses of Philippine performance in dealing with its neighbors, the author's recommendations are two-pronged: improvements of (1) defense policy, and, (2) foreign policy.

The hesitation of the national leadership in directly confronting neighboring powers that have occupied portions of Philippine territory is rooted in the weakness of the defense forces of the Philippines. On the security aspect of the author's recommendations, he is advocating that the Philippines should develop a strong defense and weapons manufacturing industry. Along this line, a military-industrial complex is a necessary component of industrialization. It will generate employment, encourage research and development, and spur scientific and technical expertise that have spillover effects in other industries. The rise of Philippine aircraft and shipbuilding industries will have far-reaching economic consequences for the entire nation. The multiplier effect of spinning off components of defense industries to civilian endeavors will redound to the comprehensive economic development of the country.

Among the aspects that defense policy has to improve on are: first, the hardware component of the armed forces. Those who are watching events in Ukraine have witnessed the role of missiles and long-range weapons in holding at bay a much bigger, visibly more powerful, and more heavily-equipped opponent. One takeaway in the Ukrainian conflict is the multidimensional infliction of damage by weapons systems that used to be defined only for certain roles. An example is the sinking of the Moskva by land-based missiles of Ukraine. There were many other vessels that were either sunk or heavily-damaged from land-based missiles or drones. The traditional definitions of land, air, and naval warfare have changed. Naval vessels, in particular, have become vulnerable to weapons on land that are within range. Naval surface warfare is no longer merely the concern of vessels that are afloat, submarines, or aircraft, but also of ground forces equipped with long-range missile systems. At the same time, the old rules still apply: tanks and armored vehicles need infantry support in areas where all-around visibility is poor, limited, or virtually non-existent. Aircraft can easily be targeted by MANPADS, or Man Portable Air Defense Systems. These lessons of warfare should not be lost on our defense planners. One consequence is that modernizing our ground forces should include equipment for active air defense, long-range artillery, missile systems, and drones. It would seem that even if the Philippines does not have many modern vessels, its land forces can be upgraded in such a way that they can complement the naval and coast guard components in case of conflict. It seems clear, as indicated by the war in Ukraine, that technical expertise of troops in all services are much needed. Gone are the days when infantry, armor, and artillery troops only have the most basic skills in weapons handling. Most weapons systems now have computer programs and their operators have to have the skills to manipulate them under fire. As technological developments exponentially become more sophisticated, counter-cyber warfare in the field can be interpreted in the context of counter-battery fire when opposing artillery units literally exchanged fire in battle.

The second aspect that needs focus is on the soft component of the security establishment. Personnel in all three services need the education, especially the technical portion of it, and the skills needed by a modern defense organization. One result of a technically-trained manpower is the attainment of a skilled citizenry, especially for those returning to civilian life. Civilian industries will have a disciplined workforce. For this reason, the author strongly advocates the full implementation of the citizen armed forces provisions of the Philippine Constitution. Section 4 of the 1987 Constitution provides that "The Armed Forces of the Philippines shall be composed of a citizen armed force which shall undergo military training and serve as may be provided by law. It shall keep a regular force necessary for the security of the State." It is crystal clear from this provision that the entire armed forces is founded on the citizen-based armed force. The regular force is a subset of the citizen armed force.

On foreign policy, realism should be made the logical analytical viewpoint of choice of Philippine diplomacy. It is a fundamental approach, it has been tested time and again, and is easy to implement. As discussed, there are only two basic components of realism: power and national interest. National interest is not difficult to determine. Power is difficult to achieve. Raw power is related to the defense industrialization proposition previously explained. But both power and national interest can be supplemented by astute and savvy diplomatic strategies. For instance, the potential of alliances has to be maximized.

The Mutual Defense Treaty (MDT) with the United States has to be maintained and updated, its mechanism for invocation refined to make it clear and "user-friendly." Its joint planning groups must formulate and master scenarios where the provisions of the Treaty can be of actual service to the alliance. The executive branches of the Philippines and the United States may wish to devise a system so that the appropriate committees of the legislatures of both countries could be informed of developments potentially falling under the MDT so that the "constitutional processes"[171] referred to in the Treaty could be fast-tracked, if necessary.

Generally speaking, there are two ways of strengthening one's country. First, one source of strength is found in the domestic setting. The enhancement of budgetary allocation, for instance, can improve national defense. Secondly, agreements and cooperative arrangements with other countries can also improve security for the nation. This is the international aspect of making the country strong. But treaties and alliances can be two-edged swords. If lacking in skilled diplomacy, a country may be dragged into a war not of its own making. Commitments can pull one ally to automatically side with its friends. The best example was Poland when it was attacked by Hitler in 1939. Great Britain had to declare war on Germany because of the former's alliance with Poland. Thus, World War Two was ignited.

Our FSO Cadets should be taught to think that the Philippines has to be able to withstand diplomatic isolation if necessary. A truly independent foreign policy means what it says: it does not rely on any country for its survival. One might say that this is easier said than done. Yes, but it can be achieved. To attain the measure of independence required, the country must be economically strong. That is the reason why economic diplomacy is one of the pillars of Philippine foreign policy. A new approach that the author advocates is the merger of all foreign-related economic units of the government into the DFA. The domestic environment shall remain in the current set up. This has been the trend in many foreign services around the world. Australia, Canada, New Zealand, to name a few, have taken this step. So far as their counterparts can tell, these three countries, among others,

---

[171] Article IV, *Mutual Defense Treaty between the Republic of the Philippines and the United States of America*, August 30, 1951, signed in Washington, D.C., on that date.

are doing very well, at least in the economic sphere of diplomatic work.

Finally, for both policies to take root over the long term, there is a need for decision-makers, as well as the bureaucracy, to realize that continuity is essential in policy crafting and implementation. It would be laudable, therefore, to put a stop to the six-year cycle of planning and execution of programs in the government, coterminous with the term of the presidency. It is not a healthy exercise to have new plans and programs all over again every time Filipinos elect a new president.

The armed forces are the instruments of foreign policy, not its master.

Hans Morgenthau, 1978.

References

ABS-CBN News, eds. *Locsin: Arbitral Award PH's contribution to boosting legal order over the seas*, ABS-CBN News, Posted at Feb 23, 2022 05:43 AM. This was a ministerial forum involving the EU and ASEAN.

Allison, Graham, and Philip Zelikow. *Essence of Decision: Explaining the Cuban Missile Crisis*, New York: Longman, 1999.

Aron, Raymond. *Peace and War: A Theory of International Relations*, Garden City, NY: Doubleday, 1966.

Atienza, Kyle Aristophere T. *Halt on WPS oil search seen to discourage investments*, Business World Online, April 21, 2022.

Bajo, Anna Felicia. *Bill declaring Philippines' maritime zones gets House's nod*, GMA News, December 6, 2021, 4:34 pm.

Barcelo, Vito. Duterte, *Xi call for restraint*, manilastandard.net, April 10, 2022, 2:00 am.

Barkin, Samuel. *Realism, Prediction and Foreign Policy*, Foreign Policy Analysis, July 2009, Vol. 5, No. 3. https://www.jstor.org/stable/2490977.

Blancaflor, MJ. *Duterte, Xi commit to boost PH-China ties amid WPS dispute*, Daily Tribune, @tribunephl_MJB, April 9, 2022, 12:29 PM.

Bondoc, Jarius. *Kahinaan ng loob, isip sinasamantala ng China*, Pilipino Star Ngayon, March 29, 2022, 12:00 am.

Bordey, Hana. *DFA Philippines filed 241 diplomatic protests vs. China since 2016*, GMA News 7:11 pm, December 6, 2021.

Brands, Hal. *American Grand Strategy and the Liberal Order: Continuity, Change, and Options for the Future*, RAND Corporation (2016). http://www.jstor.com/stable/resrep02400.

Bull, Hedley. *The Anarchical Society: A Study of Order in World Politics.* London and Basingstoke, UK: Macmillan, 1977.

Burch, Kurt. *Changing the Rules: Reconceiving Change in the Westphalian System*, International Studies Review, Summer, 2000, Vol. 2, No. 2, Continuity and Change in the Westphalian Order (Summer, 2000), Wiley on behalf of The International Studies Association https://www.jstor.org/stable/3186432.

Buszynski, Leszek. *Realism, Institutionalism, and Philippine Security,* Asian Survey, 42:3.

Buzan, Barry. *From International System to International Society: Structural Realism and Regime Theory Meet the English School,* International Organization, Summer, 1993, Vol. 47, No. 3 (Summer, 1993).

Carpenter, Ted Galen. *Many predicted NATO expansion would lead to war. Those warnings were ignored.* The Guardian.

https://ecfr.eu/article/the-long-year-top-foreign-policy-trends-for-2021/. Accessed on 28 Feb 2022 19:00 GMT.

Carr, E.H. and Miller, J.D.B. *The Realist's Realist, Reviewing the work: The Twenty Years' Crisis*, 1919-1939: An Introduction to the Study of International Relations by E. H. Carr, supra. p. 65. The National Interest, Fall 1991, No. 25 (Fall 1991), Center for the National Interest.

Caryl, Christian. *The Enigma of Mr. X*, a review of The Kennan Diaries by George F. Kennan and Frank Costigliola, The National Interest, No. 130, THE GOP'S BALANCING ACT (March/April 2014).

CGTN news. Xi Jinping at Boao 2022: Cooperation "biggest strength" in achieving a bright future for humanity, CGTN, 21 April 2022.

Checkel, Jeffrey T. Reviewing *The Constructivist Turn in International Relations Theory*, Reviewed Works: *National Interests in International Society* by Martha Finnemore; *The Culture of National Security: Norms and Identity in World Politics* by Peter Katzenstein; *Norms in International Relations: The Struggle Against Apartheid* by Audie Klotz, World Politics, Jan., 1998, Vol. 50, No. 2 (Jan., 1998), pp. 324-348, Cambridge University Press.

Dai Le and Katie Calvey, *Cultural Diversity in Politics and Media Will Create National Prosperity*, in Disruptive Asia, https://disruptiveasia.asiasociety.org/cultural-diversity-in-politics-and-media-will-create-national-prosperity.

Diesing, Paul and Snyder, Glenn H. *Conflict Among Nations: Bargaining, Decision Making and System Structure in International Crises* (Princeton, NJ: Princeton University Press, 1977).

(Lecture by the) Earl of Birkenhead, Lord Rector of Glasgow University, November 3, 1923.

Etzioni, Amitai and Ikenberry, G. John. *Point of Order: Is China More Westphalian Than the West?* Foreign Affairs, November/December 2011, Vol. 90, No. 6 (November/December 2011), Council on Foreign Relations, p. 176. https://www.jstor.org/stable/23039640.

Fearon, James and Wendt, Alexander. *Rationalism v. Constructivism: A Skeptical View,* Handbook of International Relations, SAGE (2002).

Flores, Helen. *DOE seeks Cabinet cluster OK for West Philippine Sea oil exploration*, Philstar Global, April 20, 2022.

Fromkin, David. *Remembering Hans Morgenthau*, World Policy Journal, Fall 1993, Vol. 10, No. 3 (Fall, 1993) Duke University Press.

H. L., *Spain: Foreign Relations and Policy since 1940,* Bulletin of International News, Nov. 14, 1942, Vol. 19, No. 23 (Nov. 14,

1942), pp. 1013-1018, Royal Institute of International Affairs, https://www.jstor.org/stable/25643333.

Hao, Yufan and Hou, Ying. *Chinese Foreign Policy Making: A Comparative Perspective*, Public Administration Review, Dec., 2009, Vol. 69, Supplement to Volume 69: Comparative Chinese/American Public Administration (Dec., 2009). Wiley on behalf of the American Society for Public Administration https://www.jstor.org/stable/40469084.

Hart, David K. and Scott, William G. *The Moral Nature of Man in Organizations: A Comparative Analysis*. The Academy of Management Journal, Jun., 1971, Vol. 14, No. 2 (Jun., 1971), p. 250. Academy of Management. https://www.jstor.org/stable/255310.

Hopf, Ted. *The Promise of Constructivism in International Relations Theory*,International Security, The MIT Press, Summer, 1998, Vol. 23, No.1 (Summer, 1998).

Jervis, Robert. *Hans Morgenthau, Realism, and the Scientific Study of International Politics*, Social Research, WINTER 1994, Vol. 61, No. 4, Sixtieth Anniversary 1934-1994: The Legacy of Our Past (WINTER 1994), The Johns Hopkins University Press, p. 855. https://www.jstor.org/stable/40971063.

Johnson-Bagby, Laurie M. *The Use and Abuse of Thucydides in International Relations*, International Organization, Winter, 1994, Vol. 48, No. 1 (Winter, 1994), pp. 132-133. The MIT Press. https://www.jstor.org/stable/2706917.

Johnston, P.B., Oak, G.S., and Robinson, L. *U. S.-Philippine Relations in Historical Perspective*, U. S. Special Operations Forces in the Philippines, 2001-2014. RAND Corporation, https://www.jstor.org/stable/10.7249/j.ctt1cd0md9.9.

Kamen, Henry. *The Mediterranean and the Expulsion of Spanish Jews in 1492*, Past and Present, May 1988, No. 119 (May, 1988), Oxford University Press on behalf of The Past and Present Society, https://www.jstor.org/stable/651019.

Kennan, George F. *On the Crisis Within the Soviet System*, Arms Control Today, September 1989, Vol. 19, No. 7 (September 1989).

Kennan, George F. *Long Telegram,* February 22, 1946, History and Public Policy Program Digital Archive, National Archives and Records Administration, Department of State Records (Record Group 59), Central Decimal File, 1945-1949, 861.00/2-2246; reprinted in US Department of State, ed., Foreign Relations of the United States, 1946, Volume VI, Eastern Europe; The Soviet Union (Washington, DC: United States Government Printing Office, 1969) p. 14 of declassified original document. https://digitalarchive.wilsoncenter.org/document/116178.

Knorr, Klaus and Sidney Verba, eds. *Assumptions of Rationality and Non-rationality in Models of the International System* in Klaus Knorr and Sidney Verba, eds., The International System: Theoretical Essays (Princeton, NJ: Princeton University Press, 1961).

Lalu, Gabriel Pabico. *Lacson asks DFA: What happened to 200 diplomatic protests vs China?*, INQUIRER.net / 03:38 PM December 06, 2021.

Lawrence, Gavin. *Aristotle and the Ideal Life*, The Philosophical Review, Jan., 1993, Vol. 102, No. 1 (Jan., 1993). Published by Duke University Press on behalf of Philosophical Review. Stable URL: https://www.jstor.org/stable/2185651.

Lebow, Richard Ned and Rosch, Felix. *A Contemporary Perspective on Realism*, https://www.e-ir.info/2018/02/17/a-contemporary-perspective-on-realism/ Feb. 17, 2018.

Luce, Duncan R. and Raiffa, Howard. *Games and Decisions: Introduction and Critical Survey* (New York: Wiley, 1957).

Machiavelli, Niccolo. *The Prince.* 1513.

Magallona v. The Executive Secretary, *G. R. No. 187167*, August 16, 2011.

Magcamit, Michael. *"The Duterte method: A neoclassical realist guide to understanding a small power's foreign policy and strategic behaviour in the Asia Pacific,"* Asian Journal of Comparative Politics, October, 2019.

Manila Times, eds. *Locsin meets India's External Affairs Minister*, Manila Times, February 21, 2022, https://www.manilatimes.net/2022/02/21/expats-diplomats/locsin-meets-indias-external-affairs-minister/1833618.

McCormick, John P. Social Research, Vol. 81, No. 1, *Machiavelli's The Prince 500 Years Later* (SPRING 2014], The Johns Hopkins University Press. p. xxiv.

McGinnis, John O. *Constitutional Review by the Executive in Foreign Affairs and War Powers: A Consequence of Rational Choice in the Separation of Powers,* Law and *Contemporary Problems*, Autumn, 1993, Vol. 56, No. 4, *Elected Branch Influences in Constitutional Decisionmaking* (Autumn, 1993). THE FEDERALIST No. 51, James Madison, (Clinton Rossiter ed., 1961).

Mearsheimer, John. *Why the West is principally responsible for the Ukrainian crisis.* The Economist, economist.com. Accessed on March 11, 2022.

Meiser, Jeffrey W. *Introducing Liberalism in International Relations Theory*, February 18, 2018. https://www.e-ir.info/2018/02/18/introducing-liberalism-in-international-relations-theory.

Mercado, Neil Arwin. *Sampaguita gas drilling will 'forcefully assert' PH rights over WPS—solon,* INQUIRER.net/01:20 AM, December 07, 2021.

Montevideo Convention on the Rights and Duties of States, Article 3, Montevideo. Entered into force on December 26, 1934.

https://www.jus.uio.no/english/services/library/treaties/01/1-02/rights-duties-states.xml.

Moravcsik, Andrew. *Taking Preferences Seriously: A Liberal Theory of International Politics*, International Organization, Autumn, 1997, Vol. 51, No. 4.

Morgenthau, Hans J. *Politics Among Nations: The Struggle for Power and Peace*, 5th ed.; Revised; (New York: Alfred A. Knopf, 1978).

Morgenthau, Hans J. *Realism in International Politics*, Naval War College Review, Vol. 10, No. 5 (January, 1958), p. 1. Published by the U. S. Naval War College Press. https://www.jstor.org/stable/44640810.

Mutual Defense Treaty between the Republic of the Philippines and the United States of America, August 30, 1951, signed at Washington, D.C.

Nodalo, Tristan. *Chinese envoy hopes the next PH leader will continue "friendly policy," keep momentum of PH-China ties*. CNN Philippines, April 12, 2022.

Noriega, Richa. *Philippines, Poland firm sign deal for 32 Black Hawk helicopters*, GMA News, Published February 22, 2022, 4:53 pm.

Nuechterlein, Donald E. *National interest and foreign policy: a conceptual framework for analysis and decision-making,* published online on 26 October 2009 by Cambridge University Press.

ODNI, eds. *Annual Threat Assessment of the U. S. Intelligence Community*, Office of the Director of National Intelligence, February 7, 2022.

Perkins, Dexter. *Fundamental Principles of American Foreign Policy*, The Annals of the American Academy of Political and Social Science, Nov., 1941, Vol. 218, Public Policy in a World at War (Nov., 1941).

Permanent Court of Arbitration (PCA) Press Release, "Arbitration between the Republic of the Philippines and the People's Republic of China: Arbitral Tribunal Establishes Rules of Procedure and Initial Timetable," August 27, 2013. https://pcacases.com/web/sendAttach/227. Annex VII refers to the provisions on arbitration annexed to the UNCLOS that parties may refer to for procedural requirements, jurisdiction, and grounds for institution of proceedings.

PCA Press Release, "The Tribunal Renders Award on Jurisdiction and Admissibility; Will Hold Further Hearings," October 29, 2015. https://pcacases.com/web/sendAttach/1503.

PCA Press Release, Award on the South China Sea Arbitration, July 12, 2016. https://pcacases.com/web/sendAttach/2086.

Philippine Daily Inquirer eds., (Editorial) *Power in the wrong hands*, Philippine Daily Inquirer/04:08 AM, November 26, 2019.

Pye, Lucien. *China: Erratic State, Frustrated Society*, Foreign Affairs, 69:4 (Fall 1990) in Yongjin Zhang, *System, empire and state in Chinese international relations, Review of International Studies*, Vol. 27, *Special Issue: Empires, Systems and States: Great Transformations in International Politics* (December 2001). Cambridge University Press. https://www.jstor.org/stable/45299504.

Radasanu, Andrea. *Montesquieu on Ancient Greek Foreign Relations: Toward National Self-Interest and International Peace*, Political Research Quarterly, MARCH 2013, Vol. 66, No. 1 (MARCH 2013), p. 3 (Abstract). Sage Publications, Inc. on behalf of the University of Utah. https://www.jstor.org/stable/23563585.

Rahmat, Ridzwan. *Philippines to deploy BrahMos missiles with new coastal defence unit*, janes.com, 17 January 2022.

Read, James H. *Polity*. Summer 1991, Vol. 23, No. 4 (Summer, 1991), p. 506. The University of Chicago Press on behalf of the Northeastern Political Science Association.

Reus-Smit, Christian. The Moral Purpose of the State: Culture, Social Identity, and Institutional Rationality in International Relations (Princeton, NJ: Princeton University Press, 1999).

Ringmar, Erik. *History of International Relations: A Non-European Perspective*, New Edition [online], Cambridge: Open Book

Publishers, 2019, Chapter 8, European Expansion, http://books.openedition.org/obp/9111, ISBN: 9781783740246.

Romero, Paolo. HEADLINES, *Diplomatic protests still effective vs China—experts*, The Philippine Star, December 7, 2021 l 12:00 am.

Rosenau, James. "Pre-theories and Theories of Foreign Policy." In *Approaches to Comparative and International Politics*. Edited by R. Barry Farrell, 27-92. Evanston, IL: Northwestern University Press, 1966.

Roshwald, Mordecai. *Realism and Idealism in Politics*, Social Science, Published by Pi Gamma Mu, International Honor Society in Social Sciences, April 1971, Vol. 46, No. 2.

Samonte, Mauro Gia. *What is the Xi-Duterte deterrence for war?*, The Manila Times, April 10, 2022.

Severino, Rodolfo C. *The Philippines' National Territory*, Southeast Asian Affairs, (2012). Published by ISEAS–Yusof Ishak Institute, https://www.jstor.org/stable/41713998.

Sorensen, Camilla T. N. Is China Becoming More Aggressive? A Neoclassical Realist Analysis, *Asian Perspective*, July-Sept 2013, Vol. 37, No. 3 (July-Sept. 2013). https://www.jstor.org/stable 42704834.

Spindler, Manuela. *Neorealist Theory*, International Relations: A Self-Study Guide to Theory, Published by: Verlag Barbara Budrich, 2013. http://www.jstor.com/stable/j.ctvdf09vd.8.

Tan, Alyssa Nicole O. and Tan, Jasper Emerald G. *Analysts decry unclear foreign policy direction*, Business World, Accessed on March 6, 2022.

Theys, Sarina. *Introducing Constructivism in International Relations Theory*, February 23, 2018. https://www.e-ir.info/2018/02/23/introducing-constructivism-in-international-relations-theory/.

The 1987 Constitution of the Philippines.

The Vienna Convention on Diplomatic Relations, 18 April 1961.

The Vienna Convention on Consular Relations, 24 April 1963.

United Nations Convention on the Law of the Sea (UNCLOS), 1982.

Walt, Stephen. *The World Wants You to Think Like a Realist*, Foreign Policy, May 30, 2018.

Waltz, Kenneth H. *Man, the State and War: A Theoretical Analysis.* New York: Columbia University Press, 1959.

Waltz, Kenneth N. *Structural Realism after the Cold War*, International Security, Summer, 2000, Vol. 25, No. 1 (Summer, 2000), The MIT Press. https://www.jstor.org/stable/2626772.

Waltz, Kenneth N. *"The Origins of War in Neorealist Theory,"* Journal of Interdisciplinary History, Vol. 18, No. 4, (Spring 1988).

Waltz, Kenneth H. *Theory of International Politics,* 1979.

Weforum online article. https://weforum.org/agenda/2016/04/4-trends-that-will-define-future–of-us-foreign-policy/.

Zagare, Frank C. *Rationality and Deterrence,* World Politics, Jan., 1990, Vol. 42, No.2 (Jan., 1990), Cambridge University Press. https://www.jstor.org/stable/2010465.

Zhang, Yongjin. *System, empire and state in Chinese international relations.* Review of International Studies, Vol. 27, Special Issue: Empires, Systems and States: Great Transformations in International Politics (December 2001), p. 63, Published by: Cambridge University Press. Stable URL: https://www.jstor.org/stable/45299504.

Ambassador Generoso D.G. Calonge

Ambassador Calonge is a career diplomat whose service in the Department of Foreign Affairs spans almost 38 years. He has served in Philippine diplomatic missions, five of them as Head of Post, in the United States, Russia and United Arab Emirates, as well as in several offices in Manila.

Ambassador Calonge served in Israel for three years before his assignment to Chicago as Consul General in the summer of 2014. He completed his tour of duty abroad in late January 2018. Before assignment in Baghdad as Ambassador of the Philippines to Iraq, he served as Assistant Secretary for Maritime and Ocean Affairs in Manila. In 2003, he became the first Philippine Consul General in Dubai.

He earned his bachelor's degree in economics from the University of the Philippines in 1976, completed his degree in Law, also from the University of the Philippines in 1985, and then moved on to finish his Master of Laws from Harvard Law School in 1988. In 2022, just before retirement, he obtained a Ph. D. in Development Administration, Major in Public Governance, from the Philippine Christian University in Manila.

Ambassador Calonge served as an officer in the Philippine Army for six years before joining the Philippine Foreign Service. He topped the Career Minister Examinations in 1998. He graduated from the Australian Army's Officer Cadet School at Portsea in 1979. He placed number one in the Philippine Army's Scout Ranger Class No. 33.

Ambassador Calonge holds the rank of Commodore in the Philippine Coast Guard Auxiliary. He has a reserve commission in the Philippine Army as a Lieutenant Colonel.

Ambassador Calonge is very active in Knights of Rizal affairs. He was instrumental in the founding of three (3) KOR

Chapters: Department of Foreign Affairs Pasay City Foreign Service Chapter, Philippine Embassy Washington DC Chapter (inactive), and Philippine Consulate General Chicago Maynilad Chapter. He was the founding Chapter Commander of the PCG Chicago Maynilad Chapter.

Ambassador Calonge is married to Gloria Salazar Calonge, Esq., and they have three adult children, namely, Golda, Nikki and Joshua Benedict. He has one granddaughter, Corie, daughter of Golda and son-in-law Aaron James.